# Improved Detection Methods for *E. coli* O157:H7

Awwa Research Foundation

## About the Awwa Research Foundation

The Awwa Research Foundation (AwwaRF) is a member-supported, international, nonprofit organization that sponsors research to enable water utilities, public health agencies, and other professionals to provide safe and affordable drinking water to consumers.

The Foundation's mission is to advance the science of water to improve the quality of life. To achieve this mission, the Foundation sponsors studies on all aspects of drinking water, including supply and resources, treatment, monitoring and analysis, distribution, management, and health effects. Funding for research is provided primarily by subscription payments from approximately 1,000 utilities, consulting firms, and manufacturers in North America and abroad. Additional funding comes from collaborative partnerships with other national and international organizations, allowing for resources to be leveraged, expertise to be shared, and broad-based knowledge to be developed and disseminated. Government funding serves as a third source of research dollars.

From its headquarters in Denver, Colorado, the Foundation's staff directs and supports the efforts of more than 800 volunteers who serve on the board of trustees and various committees. These volunteers represent many facets of the water industry, and contribute their expertise to select and monitor research studies that benefit the entire drinking water community.

The results of research are disseminated through a number of channels, including reports, the Web site, conferences, and periodicals.

For subscribers, the Foundation serves as a cooperative program in which water suppliers unite to pool their resources. By applying Foundation research findings, these water suppliers can save substantial costs and stay on the leading edge of drinking water science and technology. Since its inception, AwwaRF has supplied the water community with more than $300 million in applied research.

More information about the Foundation and how to become a subscriber is available on the Web at **www.awwarf.org**.

# Improved Detection Methods for *E. coli* O157:H7

Prepared by:
**Zia Bukhari** and **Janice Weihe**
American Water
1115 South Illinois Street, Belleville, IL 62220
and
**Mark W. LeChevallier**
American Water
1025 Laurel Oak Rd., P.O. Box 1770, Voorhees, NJ 08043

Sponsored by:
**Awwa Research Foundation**
6666 West Quincy Avenue, Denver, CO 80235-3098

Published by:

Awwa Research Foundation
Advancing the Science of Water®

**DISCLAIMER**

# CONTENTS

# TABLES

# FIGURES

# FOREWORD

The Awwa Research Foundation is a nonprofit corporation that is dedicated to the implementation of a research effort to help utilities respond to regulatory requirements and traditional high-priority concerns of the industry. The research agenda is developed through a process of consultation with subscribers and drinking water professionals. Under the umbrella of a Strategic Research Plan, the Research Advisory Council prioritizes the suggested projects based upon current and future needs, applicability, and past work; the recommendations are forwarded to the Board of Trustees for final selection. The Foundation also sponsors research projects through the unsolicited proposal process; the Collaborative Research, Research Applications, and Tailored Collaboration programs; and various joint research efforts with organizations such as the U.S. Environmental Protection Agency, the U.S. Bureau of Reclamation, and the Association of California Water Agencies.

This publication is a result of one of these sponsored studies, and it is hoped that its findings will be applied in communities throughout the world. The following report serves not only as a means of communicating the results of the water industry's centralized research program but also as a tool to enlist the further support of the nonmember utilities and individuals.

Projects are managed closely from their inception to the final report by the Foundation's staff and large cadre of volunteers who willingly contribute their time and expertise. The Foundation serves a planning and management function and awards contracts to other institutions such as water utilities, universities, and engineering firms. The funding for this research effort comes primarily from the Subscription Program, through which water utilities subscribe to the research program and make an annual payment proportionate to the volume of water they deliver, and consultants and manufacturers subscribe based on their annual billings. The program offers a cost-effective and fair method for funding research in the public interest.

A broad spectrum of water supply issues is addressed by the Foundation's research agenda: resources, treatment and operations, distribution and storage, water quality and analysis, toxicology, economics, and management. The ultimate purpose of the coordinated effort is to assist water suppliers to provide the highest possible quality of water economically and reliably. The true benefits are realized when the results are implemented at the utility level. The foundation's trustees are pleased to offer this publication as a contribution toward that end.

Walter J. Bishop
Chair, Board of Trustees
Awwa Research Foundation

James F. Manwaring, P.E.
Executive Director
Awwa Research Foundation

# ACKNOWLEDGMENTS

All developmental work for this project was performed at the American Water Quality Control and Research laboratories. The authors express their appreciation of the support provided by various staff in this laboratory, particularly Felicia Abrams and Rita Scanlon for their technical assistance.

The authors acknowledge the help of all American Water subsidiary utilities, particularly Brent Gregory, Lori Horstman, Esther Dundore, Tim Elliot, Thomas Chinske, Terrence Fitzgerald and Timothy Lantz from Illinois American; Bruce Manning, Steve Johnson and Paul Keck from Missouri American; Judy Starcevich, Susan Haack and Joel Mohr from Iowa American; Robin Casale, Leslie Jordon and Carla Marci from California American; Barbara Allison and Paul Zielinski from Pennsylvania American; Kevin Dixon, John Spangler and Ken Christy from New Jersey American and Tom Holbrook and Teri Merrifield from West Virginia American.

Myron Honda, Lori Suyama and Wanda Chang from Hawaii Department of Health; Stanley States from Pittsburgh Water and Sewer Authority; Mark Cosby from North Carolina Department of Agriculture and Consumer Services; Robert Donofrio, Casey Coy and Erin Natvig from NSF International and Mike Grob and Gloria Anderson from Minnesota Valley Testing Labs, Inc. are also thanked for their expert help during the multi-laboratory round robin studies.

The advice of the Project Advisory Committee (PAC)-Charles Bowling, Chicago Department of Water; Kristen Brenner, U. S. EPA; Peter Feng, U. S. FDA; Nancy Strockbine, CDC and the help of the AwwaRF Project Manager, Jennifer Warner are greatly appreciated.

# EXECUTIVE SUMMARY

Enterohemorragic *Escherichia coli* (EHEC) O157:H7, a causative agent for hemolytic uremic syndrome, has received a great deal of attention within the medical and public communities. The disease can be life threatening in the young or the elderly and numerous outbreaks of human disease have been reported where *E. coli* O157:H7 transmission occurred following consumption of undercooked meat products, especially ground beef. Transmission may also occur by person to person, from consumption of contaminated vegetables or unpasteurized juice and dairy products and by the waterborne route which has also led to outbreak situations. In the United States, *E. coli* O157:H7 has become a major public health concern with an estimated 40,000 cases and > 250 deaths occurring annually. These increasing incidences of *E. coli* O157:H7 create an urgent need for improved epidemiological surveillance, which requires the need for simple, rapid and sensitive isolation and detection procedures. Traditionally, *E. coli* O157:H7 biochemical characteristics (e.g., inability to ferment sorbitol) have been incorporated into culture procedures using sorbitol-MacConkey agar, however these approaches are tedious, time consuming, and, most importantly, lack the specificity for accurate determination.

Four immunological assays (Reveal®, Eclipse™, ImmunoCard® STAT! and *E. coli* O157 Antigen Detection Test) and one molecular assay (BAX®) were evaluated to determine their detection sensitivity for *E. coli* O157:H7. Of the immunological detection procedures, the Reveal® test was the most promising, with a sensitivity level of $1 \times 10^4$ colony forming units (CFU). Evaluation of a molecular detection procedure (BAX® *E. coli* O157:H7 assay) showed a sensitivity level of 10 CFU.

The two most promising detection procedures (one immunological and one molecular) were incorporated into an overall method consisting of 200 mL water sample concentration (membrane filtration), enrichment and simultaneous capture using immunomagnetic separation. The sensitivity of this overall procedure was determined using various spiked environmental water matrices in both single laboratory and multi-laboratory round robin evaluations to identify procedural issues as well as method robustness. The overall method sensitivity ranged between 0.03 – 0.12 CFU per mL, making the procedures adequately sensitive for environmental screening. Multi-lab evaluation determined that the Reveal® assay was more sensitive and reproducible when using spiked finished water samples, however, the BAX® method may have greater specificity; an important factor when evaluating raw water samples. While the overall methods successfully and rapidly (single day) detected the presence of *E. coli* O157:H7 in various water matrices, cross reactivity and sensitivity issues need further study.

## RESEARCH OBJECTIVES

The specific objectives of this study were to 1) determine sensitivity and specificity of proposed detection procedures for *E. coli* O157:H7; 2) determine compatibility of selected detection procedures with proposed isolation procedures for recovery of *E. coli* O157:H7 from source and finished water samples; and 3) evaluate the overall procedures in multiple utility laboratories to identify method(s) sensitivity and robustness.

In order to achieve these specific objectives, the project was divided into five tasks. The tasks were as follows:

- Task 1: Compare sensitivity of detection procedures using *E. coli* O157:H7.
- Task 2: Identify cross-reactivity issues and detection capacity for various isolates of *E. coli* O157:H7.
- Task 3: Utilize up to three sensitive/specific detection procedures with proposed isolation procedures to determine overall assay sensitivity.
- Task 4: Determine effects of source and finished water matrices on method(s) detection limits.
- Task 5: Conduct multi-laboratory validation for at least one procedure.

## APPROACH

In concert with the methods currently used by the water industry to comply with the Total Coliform Rule, several different approaches were selected for enrichment and isolation of *E. coli* O157:H7 from source and finished water concentrates. The project examined the sensitivity of four immunological lateral diffusion detection kits and a single molecular kit directly by using various spike doses of a known isolate of *E. coli* O157:H7. This enabled determination of the sensitivity thresholds of currently available detection methods that are rapid, inexpensive and user-friendly. It was considered important to identify limitations of these procedures before developing a strategy for an overall method that would provide the necessary sensitivity to detect low levels of environmental contamination with *E. coli* O157:H7. During these initial evaluations of the detection procedures the specificity of the test kits was assessed using high concentrations of non-target organisms. The investigators recognize that environmental samples can contain a myriad of non-target organisms, however, it would not only be labor intensive, but also cost prohibitive to conduct a thorough evaluation of all the non-target organisms for the cross-reactivity studies. Therefore a select group of five organisms was used for this evaluation. Once sensitivity thresholds were defined for the detection procedures, theoretical extrapolations were used from the detection method sensitivity to ascertain the subsequent steps necessary to meet the project objectives.

Sensitivity measurements of the detection methods indicated that these procedures alone would not be useful for direct detection of *E. coli* O157:H7 contamination in environmental water samples and, as a result, enrichment and/or concentration procedures were required to increase the relative sample volumes that were analyzed by the detection procedures. Multiple approaches were taken to optimize and refine these enrichment and concentrations steps. Both membrane filtration and centrifugation were compared for concentrating large sample volumes. In addition to Tryptic Soy Broth (TSB), lauryl tryptose broth, brilliant green bile broth and Colilert® media were examined for their use as enrichment media for *E. coli* O157:H7. Also various other media, including some specifically designed for *E. coli* O157:H7 enrichment, were evaluated using various incubation times and temperatures, to determine growth dynamics of *E. coli* O157:H7.

Since *E. coli* O157:H7 would likely be present in low numbers in environmental samples, the use of immunomagnetic separation (IMS) to aid in organism isolation was also assessed. Two types of IMS beads were evaluated: Dynabeads® anti-*E. coli* O157 (Dynal Inc., Lake Success, NY) and *E. coli* O157 A-Beads™ (Aureon Biosciences, Vienna, Austria). Both consist of paramagnetic beads, with affinity purified antibodies against *E. coli* O157 covalently bound to their surface, and enable specific capture of the target organisms from contaminating debris. Determination of the relative performance of these different IMS beads enabled selection of the most efficient beads for further evaluation in studies examining recovery efficiencies in larger volumes and/or elevated incubation temperatures than those originally recommended by the manufacturer.

In addition to data from controlled spiking studies using various permutations of the methods, theoretical calculations were used to determine the minimum original sample volume to achieve the targeted method sensitivity levels. With increasing sample volume, it was anticipated that one or more sample concentration procedures would be necessary to make the sample compatible with the downstream procedures. Two commonly used procedures that are accessible to water utility or research facilities are membrane filtration procedures and centrifugation. Optimization of these procedures and their incorporation into the overall method was followed by additional sensitivity and specificity measurements to ensure the overall procedures were sensitive, rapid, user-friendly and capable of yielding complete analysis within a single day. Where necessary, further tweaking of the procedures was performed. Following these method development or optimization studies, one or more procedures were evaluated in the investigators laboratory, using various initial spike doses of *E. coli* O157:H7 to ensure the overall project objectives were met. Following these studies, standard operating procedures were prepared for each developed method and circulated to independent participating laboratories where 'blind' round robin studies were conducted using spike doses of *E. coli* O157:H7 provided by the investigators laboratory. The round-robin approach was intended to determine whether laboratories could perform the methods developed/optimized during this research without the need for extensive initial training (*i.e.*, assess how user-friendly these procedures were). More importantly, the objective was to ascertain information on the sensitivity, specificity and robustness of the overall procedures.

## CONCLUSIONS

The overall conclusion from this research is that two detection procedures (Reveal® and BAX®), when combined with an optimized concentration, enrichment, and isolation procedure, enable rapid (8 h) detection of *E. coli* O157:H7 from source or finished water samples.

1. One immunological (Reveal® assay) and one molecular assay (BAX® assay) allowed reproducible detection of $1 \times 10^4$ CFU and 10 CFU per test, respectively, during the sensitivity testing of the detection procedures.

2. Tryptic Soy Broth (TSB) was the most effective enrichment medium for growth of *E. coli* O157:H7. The propagation rate indicated temperature dependency, with 42°C yielding faster growth rates than 35°C.

3. The Dynal immunomagnetic separation (IMS) beads performed better than Aureon when using 1 mL sample volumes. Scaling up the IMS to 10 mL samples had no impact on *E.*

*coli* O157:H7 recoveries. Reducing the Dynal bead concentration by 50% (i.e. 200 μL to 100 μL) in scaled up IMS demonstrated a marginal reduction in recoveries but was a cost-effective option.

4. Spiking *E. coli* O157:H7 alone, or in combination with non-target organisms, in raw or finished water concentrates followed by IMS indicated no significant reduction in recovery rates. This indicated that the water matrix or the competing organisms examined did not have a significant deleterious effect on the IMS performance. The non-target organisms had less than a 2% carry-over, indicating the antibodies on IMS beads had low cross reactivity for these non-target organisms.

5. *E. coli* O157:H7 spiked into reagent-grade (Milli-Q) showed membrane filtration yielded higher recoveries than centrifugation. Addition of TSB (single strength) increased the effectiveness of centrifugation; however, membrane filtration was selected for further trials due to the ease of use, and lower cost.

6. For both the Reveal® and BAX® methods, multi-laboratory trials validated the single laboratory data on method sensitivity and reproducibility. Over-all procedures had a fast turn-around time (<8 h) and were sufficiently sensitivity for environmental monitoring (esp. emergency situations) for *E. coli* O157:H7.

7. The Reveal® test was inexpensive ($23.82) and required no other specialized equipment. Although the cost of consumables for the BAX® assay was inexpensive ($24.42), the test required the use of proprietary equipment that cost approximately $25,000.

## RECOMMENDATIONS AND FUTURE RESEARCH

1. Methods from this project should enable water utilities to adopt a two-prong approach for analysis of *E. coli* O157:H7. Following a total coliform test, or *E. coli* analysis, methods described here could be applied to rapidly determine presence of *E. coli* O157:H7 directly in a sub-sample of the total coliform or *E. coli* positive samples. Alternatively, these methods could be applied to test directly for *E. coli* O157:H7 contamination in up to 200 mL sample volumes of source or finished water.

2. The described procedures are rapid and user-friendly and their application as preliminary tools in epidemiological investigations of outbreak situations is recommended. This will allow employment of rapid contingency measures, where necessary, to eliminate or reduce the potential health risk to the public.

3. Using geographically distinct sources, occurrence of *E. coli* O157:H7 in source and finished water samples and information on seasonal distribution is feasible with these methods.

4. These methods may also be employed for assessing the efficiency of water treatment processes for removing and/or inactivating *E. coli* O157:H7.

5. Caution should be exercised in interpretation of data from the immunological assay when used for raw water samples (potential cross reactivity issues). It is recommended that

extensive trials are conducted to determine the relative cross reactivity of the Reveal® and BAX® assays using raw water samples taken from various sources.

6. In finished water, both assays have application as preliminary monitoring tools, but confirmation of presumptive positives will remain imperative.

# CHAPTER 1
# INTRODUCTION AND BACKGROUND

*Escherichia coli* are facultative anaerobes that commonly inhabit the gastrointestinal tract of humans within hours of birth. This gram negative bacillus of the family *Enterobactericeae* is normally a harmless commensal of the human body, which can aid the host by suppressing pathogenic bacteria as well as producing vitamin K and vitamin B complex (Weir, 2000). In individuals with severe immunosuppression or other conditions that compromise the gastro-intestinal barriers, there can be cases of infection with non-pathogenic *E. coli*. More likely are instances of infection with pathogenic *E. coli,* which can include urinary tract infections, sepsis/meningitis or diarrheal disease. Within the pathogenic *E. coli* responsible for diarrheagenic disease, differentiation of the strains has lead to the development of five classes: enteropathogenic *E. coli* (EPEC), enterohemorrhagic *E. coli* (EHEC), enterotoxigenic *E. coli* (ETEC), enteroinvasive *E. coli* (EIEC) and enteroaggregative *E. coli* (EAEC). Differentiation of pathogenic strains of *E. coli* is based on the organism's surface antigen profiles of O (somatic), H (flagellar) and K (capsular). Kaufmann originally proposed serological classification in 1944 and a modified form of this is still used today.

Investigation of two outbreaks of gastrointestinal disease with individuals experiencing severe cramping, abdominal pains, watery diarrhea leading to bloody diarrhea and low-grade fever, resulted in identification of EHEC as a distinct class of pathogenic *E. coli* (Riley *et al.,* 1983). The illness, which was designated hemorrhagic colitis (HC), was associated with consumption of undercooked hamburgers from a fast food restaurant and examination of stool cultures from affected individuals revealed the presence of a rare *E. coli* serotype, O157:H7. Also in 1983 sporadic cases of hemolytic uremic syndrome (HUS) were reported by Karmali *et al.*, (1983b) with detection of fecal cytotoxins and cytotoxin producing *E. coli* in stool samples. Karmali *et al.,* (1983a) reported that Verocytotoxin/Shiga like toxins were the common virulence factor between HC and HUS, which led to recognition of *E. coli* O157:H7 as an emerging enteric pathogen that was responsible for both intestinal and renal disease. Whether the organism was really an emerging pathogen was confirmed by retrospective studies conducted by the Centers for Disease Control (CDC) and Public Health Laboratory Services (PHLS) in the United Kingdom, which indicated the presence of *E. coli* O157:H7 in one of 3,000 samples and one of 15,000 samples respectively. It has been suggested that the reason for this emergence was the acquisition of virulence factors, such as the Shiga toxin gene, by horizontal gene transfer and recombination (Mead and Griffin, 1998). *E. coli* O157:H7 produces Shiga toxins, thus allowing it to be categorized amongst other Shiga toxin producing *E. coli* (STEC), alternatively called Verocytotoxin producing *E. coli* (VTEC). Both terms are synonymous and are seen in the literature interchangeably. Various strains of *E. coli* O157:H7 can produce Shiga toxin 1 (Stx1) or Shiga toxin 2 (Stx2) or both. Stx1 is virtually identical to the toxin produced by *Shigella dysenteriae* type I, while Stx2 shares approximately 56% homology to Stx1 (Mead and Griffin, 1998).

## CLINICAL FEATURES AND PATHOGENESIS OF *E. COLI* O157:H7

Infection with *E. coli* O157:H7 brings with it a wide range of clinical manifestations ranging from asymptomatic carriage to HC and HUS. The incubation period between exposure and disease onset ranges from one to eight days. The illness typically presents with abdominal cramps and non-bloody diarrhea, which progress to bloody diarrhea after one to two days in 70%

of cases. There may only be a small amount of blood in the stool or stools may be almost entirely blood (Coia, 1998). Fever occurs in only 30% of patients with vomiting occurring in 30-60% (Mead and Griffin, 1998). While most cases of HC resolve themselves within one week, other more severe cases may progress to HUS. The percentage of HC cases that advance to HUS ranges from 3-20%. Factors that may contribute to this progression include fever, age-related susceptibility (i.e., children and the elderly being more susceptible), elevated serum leukocyte count, and treatment with antimotility agents (Mead and Griffin, 1998). The role of antibiotic treatment in the progression to HUS remains controversial. Conflicting studies have been reported that show both increased and decreased levels of HUS among patients receiving antibiotic therapy. Death can occur in 3-5% of individuals with HUS, while 12-30% may suffer a variety of long term renal damage (Nataro and Kaper, 1998). In some patients, particularly the elderly, the disease also progresses to thrombotic thrombocytopenic purpura (TTP), causing neurological deficits and involvement of organs other than the kidneys (Weir, 2000).

The mechanisms by which *E. coli* O157:H7 cause HC and HUS has been extensively investigated. The organism adheres to mucosal cells in the gastrointestinal tract, disrupting the brush border and causing characteristic attachment and effacing lesions. Although the Shiga toxins produced by O157 are the most frequently discussed virulence factors, they are not the sole cause of damage. Presence of a 60 MDa plasmid, which encodes for a enterohaemolysin, is thought to cause lysis of erythrocytes, as is the case with bloody diarrhea from HC, leading to the release of heme and hemoglobin that may subsequently be used as an iron source for O157 (Nataro and Kaper, 1998; Mead and Griffin, 1998). A second virulence factor important to O157 pathogenicity is the locus of enterocyte effacement (LEE), which contains the *eae* and *tir* genes. The *eae* gene encodes for intimin, a bacterial cell surface protein, which along with a translocated intimin receptor encoded by *tir*, allows for adherence of *E. coli* O157:H7 to the intestinal mucosa. This produces the attachment and effacing lesions seen in *E. coli* O157:H7 infections (Coia, 1998; Ibekwe *et al.*, 2002).

*E. coli* O157:H7 can produce Stx1 or Stx2 or both, with studies showing that Stx2 causes greater damage to human renal cells (Li and Drake, 2001) and is responsible for a larger number of HUS cases (Ochoa and Cleary, 2003). The shiga toxins are bacteriophage encoded and consist of five B subunits and one A subunit. The B subunits bind to globotriaosylceramide (Gb3), a glycolipid found in high levels in the human kidney (Kehl, 2002). Once internalized within the cell the A subunit inactivates the 28S rRNA of the 60S ribosomal subunit thereby preventing protein synthesis and resulting in cell death (Mead and Griffin, 1998). It is believed that this process is the primary cause of renal cell damage in HUS (Coia, 1998).

## EPIDEMIOLOGY

The Centers for Disease Control (CDC) estimates the annual disease burden from *E. coli* O157:H7 to be around 40,000, with up to 250 deaths (Boyce, Swerdlow, and Griffin 1995), however, the true incidence may be under-estimated due to failure of laboratories to successfully isolate or detect this pathogen. Other researchers have estimated the number of incidences to be as high as 73,000 annually with 11,000 of those being from waterborne transmission (Shelton and Karns, 2001). Interestingly the disease has become of greater importance in developed countries of the northern hemisphere, where there has been some association with seasonality, and higher incidence in the summer months. In contrast, *E. coli* strains other than *E. coli* O157:H7 have been responsible for diarrhea in developing countries and this has occurred despite concerted efforts to seek this organism in a number of studies (Cravioto *et al.*, 1990; Albert *et al.*, 1995).

A wide variation has been noted in the shedding patterns of *E. coli* O157:H7 in human infections. In one study, 66% of stool samples were negative for *E. coli* O157:H7 seven days after diarrhea in the absence of antibiotic therapy (Tarr *et al.*, 1990). Other studies have noted a median shedding period of 17 days (range: 2 – 62 days) (Belongia *et al.*, 1993) and 21 days (range 5 – 124 days) (Karch *et al.*, 1995). In two patients changes were noted in the *E. coli* profile over two weeks of shedding, suggesting a loss in the Stx gene. This has important implications for molecular techniques for distinguishing this organism in epidemiological studies.

Although Stx producing *E. coli* can be found in the flora of various animals (cattle, sheep, pigs, cats, dogs) and birds (chicken, gulls), most are strains other than O157:H7 and their pathogenicity to man may be unclear. Cattle are recognized as important animal species responsible for harboring Stx positive *E. coli*, with infection rates typically ranging from 10% to 25%, however incidence rates up to 60% have been identified. Isolation rates of non-*E. coli* O157:H7 serotypes are considerably higher than *E. coli* O157:H7 (0-2.8%) and generally younger animals tend to have a higher occurrence of infection than older animals (Nataro and Kaper, 1998).

## TRANSMISSION ROUTES AND DOCUMENTED WATERBORNE OUTBREAKS

Enterohemorragic *E. coli* (EHEC) can be transmitted by any route where fecal contaminated material is ingested by a susceptible host, and epidemiological investigations of outbreaks suggest the infectious dose for EHEC may be 10-200 organisms (Willshaw *et al.*, 1994). Although a number of outbreaks have been reported from consumption of undercooked hamburgers either in restaurants or within the home setting, others have suggested that although this has important implication in modern food processing practices where beef from various sources is ground together and distributed statewide, more indirect sources of contamination are also likely to occur. Examples of such indirect occurrences have included improper hygiene among food handlers or inadequate washing of hands after handling ground beef, which can lead to contamination of other produce. Further examples of indirect contamination include outbreaks where other foods (mayonnaise, unpasteurized apple juice and fermented hard salami) were implicated. Some of these food products are known to have an acidic pH, where other bacteria would not survive, which also indicates the acid tolerance of *E. coli* O157:H7. Other food associated outbreak incidents have involved lettuce, uncooked radish sprouts and alfalfa sprouts. Other vehicles of disease transmission have included outbreak incidents following consumption of yogurt in UK, which affected 16 people (Morgan *et al.*, 1993). Although *E. coli* O157:H7 is susceptible to pasteurization, this outbreak suggests potential contamination post-pasteurization. In addition, over a five month period, 15 incidents were reported among children in northern Italy with individuals suffering HUS. Although no definitive mode of transmission was identified, contact with live poultry or chicken coops was noted (Tozzi *et al.*, 1994). *E. coli* O157:H7 can rapidly colonize cecal tissue of chicks, resulting in long-term (up to 11 months) shedding of this microorganism, and subsequent contamination of the egg shells (Schoeni and Doyle, 1994). It is conceivable, therefore, that live poultry were the source of infection in the outbreaks reported from northern Italy.

In addition to foodborne infections, a number of different incidents have demonstrated transmission of *E. coli* O157:H7 by the waterborne route, with human infections occurring from consumption of well water, recreational water and even municipal water. Examples of such outbreaks are presented in Table 1.1.

The first waterborne outbreak associated with *E. coli* O157:H7 occurred in Missouri in 1989 (Swerdlow *et al*., 1992), where more than 240 people were infected resulting in the death of four individuals. Outbreak investigations revealed that the source of the outbreak was likely to be backflow during two water main breaks, which may have led to pathogen intrusion. When coupled with improper maintenance of chlorine residuals in supply water, the potential pathogen intrusion event may have exacerbated this incident. Another outbreak involving *E. coli* O157:H7 occurred in 1991, and affected 21 individuals that had swam in a recreational lake up to three weeks previously in the vicinity of Portland, Oregon (Keene *et al*., 1994). Similarly, water from a children's paddling pool was responsible for transmission of infection to two other children in Scotland, UK in 1992 (Brewster *et al*., 1994). Epidemiological investigations did not reveal conclusive information; however, it was postulated that a child with diarrhea contaminated the pool with *E. coli* O157:H7 and as the pool water was not changed or disinfected, it served as the vehicle of infection for the two children. They in turn infected four others by person-to-person contact. In Wyoming 157 persons were affected with *E. coli* O157:H7 and this was associated with a community water system supplied by a spring and two wells. The water in this system was not treated, and the outbreak may have resulted from fecal contamination by wildlife near the spring (Olsen *et al.,* 2002). An *E. coli* O157:H7 outbreak in Illinois, which affected three people, occurred as a result of drinking water from an untreated well near a cattle pasture. An outbreak also occurred in Washington at a trailer park and was associated with a chlorinated groundwater supply. Four individuals were symptomatically infected and there was one hospitalization. Although the source of contamination was not determined, the water system was out of compliance with state and county regulations (Barwick *et al.,* 2000).

In May-June of 2000, a waterborne outbreak of *E. coli* O157:H7 occurred in a small rural community of Walkerton, Ontario. The outbreak situation was first identified by a notable increase in pediatric bloody diarrhea, followed by increased diarrhea occurrence in long term care facilities, elementary schools and emergency departments in the Walkerton area. Epidemiological investigations identified 1,346 reported cases of gastroenteritis from individuals with exposure to Walkerton municipal water and the estimated overall cases associated with Walkerton exceeded 2,000. Stool sample analysis from 167 of 174 people confirmed infection with *E. coli* O157:H7 and 116 individuals had confirmed infections from *Campylobacter* species. As a result of this outbreak, 65 people were hospitalized, with 27 developing hemolytic uremic syndrome. In an announcement from the office of the chief coroner, based upon the expert committee appointed to study deaths related to the Walkerton outbreak, four of 27 deaths were directly caused by the *E. coli* outbreak in Walkerton, with *E. coli* being a contributing factor leading to 3 more deaths. The cross sectional study of Walkerton households also revealed that individuals consuming Walkerton water were 11.7 times more likely to develop gastroenteritis compared to those not exposed to Walkerton water. During the outbreak, one of the two wells used as a supply source was contaminated with coliforms and *E. coli*, which was later confirmed to be *E. coli* O157:H7. Data indicated the tendency of this well to be prone to surface contamination, especially following flooding conditions, which is what preceded the Walkerton incident. Furthermore, environmental testing of livestock farms, especially a farm adjacent to the contaminated well, indicated the presence of both *E. coli* O157:H7 and *Campylobacter* infections in livestock. (Canada Communicable Disease Report, 2000)

**Table 1.1**
**Outbreaks of *E. coli* O157:H7 due to transmission by the waterborne route**

| Year, Location | # Infected | Description | Reference |
|---|---|---|---|
| 1989, Missouri, US | 243 | 2 water main breaks may have led to backflow in an unchlorinated municipal water supply leading to 4 deaths | Swerdlow *et al,* 1992 |
| 1990, Scotland | 4 | O157 contamination of reservoir water supply | Dev, Main, and Gould 1991 |
| 1990, Japan | 42 | O157 cultured from tap and well water | Akashi *et al.*, 1994 |
| 1991, Oregon, US | 21 | Fecal contamination of recreational lake | Keene *et al.*, 1994 |
| 1992, Scotland | 6 | Child with diarrhea contaminated wading pool | Brewster *et al.,* 1994 |
| 1992, South Africa and Swaziland | 86 | Cattle feces present in river used as water source | Isaacson *et al.*, 1993 |
| 1993, Rotterdam, NL | 5 | Illness amongst swimmers in recreational lake | Cransberg *et al.*, 1996 |
| 1993, London, England | 6 | Illness amongst swimmers in wading pool | Hildebrand *et al.*, 1996 |
| 1994, New York, US | 6 | Illness amongst swimmers in recreational lake | Ackman *et al.*, 1997 |
| 1995, Illinois, US | 8 | Fecal contamination of recreational lake | Warrner *et al.*, 1996 |
| 1995, Scotland | 6 | Sewage contaminated public water supply | Jones and Roworth, 1996 |
| 1995, Minnesota, US | 33 | Contaminated spring water used as drinking water supply source at campgrounds | Levy *et al.*, 1998 |
| 1995, Minnesota, US | 8 | Contaminated recreational lake | Levy *et al.*, 1998 |
| 1995, Wisconsin, US | 8 | Contaminated recreational lake | Levy *et al.*, 1998 |
| 1996, Minnesota, US | 6 | Contaminated recreational lake | Levy *et al.*, 1998 |
| 1996, Georgia, US | 18 | Contaminated pool at mobile-home park | Levy *et al.*, 1998 |
| 1997, Finland | 4 | Illness amongst visitors to lake beach | Paunio *et al.*, 1999 |
| 1997, Washington, US | 4 | Contaminated chlorinated groundwater supply at a trailer park | Barwick *et al.*, 2000 |
| 1997, Missouri, US | 8 | Illness amongst swimmers in recreational lake | Barwick *et al.*, 2000 |
| 1998, Illinois, US | 3 | Well water contaminated by cattle feces | Barwick *et al.*, 2000 |

(continued)

**Table 1.1 (Continued)**

| Year, Location | # Infected | Description | Reference |
|---|---|---|---|
| 1998, Minnesota, US | 5 | Illness amongst swimmer in recreational lake | Barwick *et al*., 2000 |
| 1998, Wyoming, US | 157 | Contamination of a spring and two wells which were municipal water source | Olsen *et al*., 2002 |
| 1999, Texas, US | 22 | Inadequately chlorinated city well water | Lee *et al*., 2002 |
| 1999, Wisconsin, US | 5 | Illness amongst swimmers at fecal contaminated recreational lake | Lee *et al*., 2002 |
| 1999, Washington, US | 36 | O157 contamination of recreational lake water | Lee *et al*., 2002 |
| 1999, New York, US | 127 | Septic system contaminated well used for drinking water at fairgrounds, resulting in 2 deaths | New York Department of Health, 2000 |
| 2000, California, US | 5 | Fecal contamination of creek water drank by campers | Lee *et al*., 2002 |
| 2000, Utah, US | 102 | Illness amongst those drinking from an irrigation system | Lee *et al*., 2002 |
| 2000, Ohio, US | 29 | Possible back flow at fairgrounds from animal barn to food vendor taps | Lee *et al*., 2002 |
| 2000, Connecticut, US | 11 | Illness reported amongst swimmers at recreational lake | Lee *et al*., 2002 |
| 2000, Walkerton, Ontario, Canada | 167 | Improper chlorination of municipal well water contaminated with O157 from surface water runoff resulting in 7 deaths | Hrudey *et al*., 2003 O'Conner 2002 |

## ROUTINE MICROBIAL METHODS USED BY THE WATER INDUSTRY

Microbial contamination of public water systems is regulated under a number of different United States Environmental Protection Agency (USEPA) regulations such as the Safe Drinking Water Act (SDWA) of 1974, with amendments introduced in 1986 and 1996. The amendments require the USEPA to publish a candidate-contaminant list every five years. These contaminants are known or anticipated to occur in public water systems and could possibly require regulation. Microbial contaminants within this list are regulated by the Surface Water Treatment Rule and Total Coliform Rule (TCR) of 1989. Later (1998), the Interim Enhanced Surface Water Treatment Rule was promulgated for additional protection from *Cryptosporidium parvum* and other waterborne pathogens. According to EPA regulations, a system that operates at least 60 days per year, and serves 25 people or more or has 15 or more service connections, is regulated as a public water system. Under the TCR, routine monitoring of tap water is required for all public water systems, with the frequency and number of samples varying with the size of the system and the number of people that are served. The presence of total coliforms indicates that the system is contaminated or vulnerable to fecal contamination. The TCR requires subsequent analysis of these positive samples for fecal coliforms or *E. coli*, demonstrated by β-glucuronidase activity. Since *E. coli* O157:H7 lacks this enzymatic activity, its presence in total coliform positive samples can not be detected by the current analysis. For any size system, presence of two consecutive total coliform positive samples from one site, with one fecal coliform or *E. coli* positive sample, results in an acute violation of the Maximum Contaminant Level (MCL) and leads to immediate mandatory notification of the public.

Various procedures have been documented in the 20th edition of Standard Methods of the Examination of Water and Wastewater (APHA *et al.,* 1998) to determine the presence of total coliforms. A multiple tube fermentation test involves a presumptive phase consisting of tubes of lauryl tryptose broth observed for evidence of turbidity and/or gas production following incubation (35°C) for 24 to 48 h. This is followed by a confirmation phase requiring incubation (35°C) of positive samples in brilliant green lactose bile broth for further evidence of the fermentation process (gas production) within 48 h. EC broth is used for differentiation into fecal and non-fecal coliforms. Another simplified procedure for detecting the coliform group is the Presence/Absence (P/A) test, which can be used to examine fermentation reactions in 100 mL samples by incubation in commercially available P/A broth, and this approach provides opportunity for further screening of the culture for other indicators (i.e. fecal coliforms). Many labs also use the commercially available Colilert® (IDEXX Laboratories, Westbrook, Maine). Colilert® utilizes a patented Defined Substrate Technology® that detects both total coliforms and *E. coli*. For larger volumes of drinking water or natural waters (100 mL – 1L), the membrane filtration (MF) technique, as well as centrifugation, can be employed for sample concentration.

## DIAGNOSTIC PROCEDURES AND PHENOTYPIC VARIANTS

Using multilocus enzyme electrophoresis of *E. coli* strains associated with enteric disease, it has been demonstrated that the O157:H7 serotype is in a well-defined group that is distantly related to other Shiga toxin-producing serotypes (Whittam and Wilson, 1988; Whittam *et al.*, 1993). Unlike the majority of other *E. coli* isolates, most strains of *E. coli* O157:H7 do not ferment sorbitol within 24 h (Farmer and Davis, 1985). This allows for an easy identification of O157 on sorbitol containing media such as Sorbitol-MacConkey (sMAC) agar. Colonies of O157 appear colorless while sorbitol fermenters appear pink or mauve. The use of sMAC agar is one of the primary screening methods used in the clinical laboratory to analyze patient specimens for the presence of serotype O157:H7 (March and Ratnam, 1986). However, the use of sMAC agar is most successful early in the disease, and, as infection progresses, the detection rate with sMAC agar drops to 33%. It has also been reported that sMAC agar has a low sensitivity of only 50-60% (Kehl, 2002). The use of cefixime-tellurite sorbitol-MacConkey agar (Chapman *et al.*, 1991) has been shown to increase sensitivity and inhibit other sorbitol non-fermenting organisms (Coia, 1998; Kehl, 2002). *E. coli* O157:H7 does not produce β-glucuronidase and therefore is negative by the 4-methyl-umbelliferyl β-D-glucuronide (MUG) assay (Doyle and Schoeni, 1984). A variety of selective media have been formulated to utilize these two biochemical characteristics through the addition of fluorogenic or chromogenic compounds (Nataro and Kaper, 1998). Although many of these selective media are useful, isolation and identification of *E. coli* O157:H7 with these methods has limitations. One drawback is the time consuming nature of the procedures. Current USDA methods involve four enrichment and culturing steps, followed by ten biochemical confirmation tests, which requires a minimum of three days (Shelton and Karns, 2001). In addition there may be misidentification of *E. coli* O157:H7 as other intestinal bacteria (i.e., *E. hermanii* and *Hafnia* spp) share similar sorbitol non-fermenting phenotype and resemble serotype O157:H7 on sorbitol-containing medium. Also *E. coli* O157 that are not the H7 serotype and are not pathogenic are also unable to ferment sorbitol (Willshaw *et al.*, 1993). In addition, there is the issue of the selective media excluding other pathogenic *E. coli* serotypes, many of which ferment sorbitol but can also be responsible for cases of hemorrhagic colitis and HUS (Mariani-Kurkdijian *et al.*, 1993, Bockemuhl, Aleksic, and Karch 1992). This is especially true in other countries, where other STEC strains such as *E. coli* O111 and O26 are responsible for a greater number of infections than *E. coli* O157:H7, whereas in the US *E. coli* O157:H7 causes 80% of STEC infections (Ochoa and Cleary, 2003). There is also evidence to indicate that serotype O157:H7 in sorbitol-containing foods can adapt to become a sorbitol-fermenting phenotype (Fratamico, Buchanan, and Cooke 1993). Isolation of sorbitol fermenting *E. coli* O157:H7 from HUS patients has also been reported (Gunzer *et al.*, 1992), with these strains also being MUG assay positive. While these findings were initially considered atypical, increasing prevalence of sorbitol fermenting *E. coli* O157:H7 from HUS patients has been reported (Bitzen *et al.*, 1993) and suggests that *E. coli* O157:H7 is capable of slow fermentation of sorbitol.

Due to the phenotypically similar traits between organisms, as well as development of phenotypic variants of serotype O157:H7, serological confirmation with O157 and H7 antisera becomes necessary (Padhye and Doyle, 1992). Anti-O157 sera have been effective for use in agglutination or latex agglutination tests and enable rapid screening or serological confirmation of isolates. Anti-O157 antibodies have been coupled to immunomagnetic beads in order to selectively isolate this pathogen from the sample matrix (Wright, Chapman, and Siddons 1994) or used in enzyme immunoassays to directly detect serotype O157:H7. Obviously, using antibody based isolation and detection procedures can be prone to cross-reactivity issues with other organisms (i.e., *Citrobacter freundii, E. hermanii, Yersinia enterocolitica*) or other non-H7 serotypes. However these issues can be reduced by either pre-adsorption procedures or by selection of specific target antigens. Using assays specific for toxins or toxin genes does not provide sufficient data for epidemiological investigations and "trace-back" studies. Large numbers of *E. coli* serotypes (>60) have been found to produce Shiga like toxins, therefore detection of potential Shiga toxin producing strains does not confirm *E. coli* O157:H7 associated illness.

Molecular assays, using the polymerase chain reaction (PCR), are less prone to such limitations. For example, one PCR based assay amplifies an allele in the *uid*A gene that is unique only to serotype O157:H7, including phenotypic variants of serotype O157:H- that are sorbitol and 4-methyl-umbelliferyl β-D-glucuronide positive (Feng, 1993). Coupled with primers specific for *Stx* genes, a multiplex PCR assay can be used to simultaneously identify isolates of serotype O157:H7 and determine the type of Shiga toxin they encode (Cebula, Payne, and Feng 1995). Analysis of pure culture isolates demonstrated that this assay detected all Shiga toxin producing serotypes and was able to distinguish isolates of serotype O157:H7, including the phenotypic variants. Multiplex PCR assays can be used to specifically target other virulence factors, such as enterohemolysin and intimin (Kehl, 2002). Advantages of these and other molecular methods include specificity, sensitivity, and the ability to detect phenotypic variants of serotype O157:H7. However, complexity and expense of these assays may be factors that limit their application to expert research facilities rather than a routine diagnostic laboratory.

While emergence of phenotypic variants does present a concern, they have occurred sporadically and are not prevalent worldwide. Nonetheless, should the frequency of isolation or the incidence of infection caused by these variants increase or should other Shiga toxin producing serotypes of *E. coli* become established as causative agents of illness, it would be advantageous to have multiple diagnostic assays available in the arsenal for detection of these pathogens in food and environmental water samples.

## COMMERCIAL DIAGNOSTIC TESTS

The media attention received by some of the recent *E. coli* O157:H7 outbreaks and scares associated with countrywide recall on meat products by large distributors within the US, has stimulated the development of many new rapid assays to detect serotype O157:H7. Currently more than 20 tests for detection of *E. coli* O157:H7 are available commercially and the majority of these assays are antibody based, using polyclonal, pre-adsorbed, affinity purified or monoclonal antibodies targeting surface expressed antigens. These antibody based tests have been used primarily in an enzyme linked immunosorbent assay (ELISA) format. The most common assay format is the "sandwich" type ELISA, where the same antibody is used to both capture and detect the specific antigens. Antibodies have been used in other novel formats, such as the immunomagnetic separation procedure (IMS) to specifically capture the organisms onto the surface of paramagnetic beads. The IMS procedures have also been combined with ELISA, using colorimetric or fluorometric detection procedures. Simpler procedures have used lateral flow immunoassays, where the antibody-antigen complex is allowed to migrate along a membrane and is captured by a second antibody that is immobilized at a specific region of the membrane. The lateral flow format, which is similar to the pregnancy test kits, produces a visual signal (with the aid of colloidal gold or dye encapsulating liposomes) when samples are positive. These tests offer an advantage because they can be performed without the need for specialized skills or equipment. Other user-friendly assays utilize molecular detection procedures such as the BAX® system. Here, primers targeting sequences derived through random amplified polymorphic DNA (RAPD) analysis, as well as all the other PCR components, are provided in a tablet format. Simple rehydration of the tablet with the lysed test sample, followed by sample incubation in a thermal cycler (with pre-set thermocycling conditions), enables specific detection of *E. coli* O157:H7.

The goal of this research was to develop a rapid method (within one working day) for the detection of *E. coli* O157:H7 from both source and finished water matrices. It was also important that procedures were sufficiently user-friendly to be easily employed by microbiology testing laboratories within water utilities. More innovative methods currently being developed (e.g., electrochemiluminescence or flow cytometry based) were either too early in their developmental stage or required the use of highly sophisticated or expensive instruments not typically found in water utility laboratories and were excluded from this study. This study examined a number of different commercially available detection procedures, which were primarily antibody based and utilized lateral flow detection.

## DETECTION PROCEDURES SELECTED FOR EVALUATION

This project evaluated several immunological and one molecular detection procedure. These were relatively easy to perform and had the potential to be completed within eight hours. The selected commercial detection procedures are summarized in Table 1.2.

**Table 1.2**
**Commercial *E. coli* O157:H7 tests selected for use in this study**

| Test Kit | Analytical Technique | Total Test Time[1] | Costs per test | Supplier |
|---|---|---|---|---|
| BAX® for Screening/*E. coli* O157:H7 | Polymerase chain reaction | 8 h | $8.50 | Qualicon, Inc.<br>P.O. Box 80357<br>Wilmington, DE 19880-0357 |
| Eclipse™ *E. coli* O157:H7 Rapid Color Change Test | Immunoassay | 8 h | $ 10.50 | Eichrom Technologies, Inc.<br>8205 South Cass Ave., Suite 111<br>Darien, IL 60561 |
| ImmunoCard® Stat! *E. coli* O157:H7 | Immunoassay using colloidal gold | 8-24 h | $10.30 | Meridian Diagnostics<br>3471 River Hills Dr.<br>Cincinnati, OH 45244 |
| Reveal® Microbial Screening Test for *E. coli* O157:H7 | Sandwich ELISA | 8 h | $8.00 | Neogen Corporation<br>620 Lesher Pl. Lansing, MI 48912 |
| *E. coli* O157 Antigen Detection Test | Antibody-dye conjugate complex | 7 h | N/A<br>Kit no longer produced | Mionix (formerly Morningstar Diagnostics, Inc.)<br>1832 Centre Point Cir., Ste. 103<br>Naperville, IL 60563 |

[1]Includes enrichment

## Background Information on Detection Procedures

### *Eclipse™ E. Coli O157:H7 test (Eichrom Technologies, Inc, Darien,Il)*

The Eclipse™ test is an immunochromatographic device that utilizes a lateral flow method for rapid detection of *E. coli* O157:H7. Immunochromatographic devices rely on the formation of an antibody-antigen "sandwich" complex, which is allowed to migrate by liquid capillary action on a nitrocellulose membrane to a point further upstream where a secondary antibody has been immobilized. Capture and binding of the complex, by the immobilized antibody, leads to accumulation of visible color at a designated point along the nitrocellulose membrane and serves as a means of qualitatively detecting the presence of the target organism (Figure 1.1). The test also contains an internal positive control that helps to monitor whether the assay functioned appropriately during a test run.

### *ImmunoCard® Stat! E. coli O157:H7 Test (Meridian Diagnostics Inc., Cincinnati, OH)*

This assay utilizes the lateral flow concept and incorporates the use of gold particles conjugated to a monoclonal antibody specific for *E. coli* O157 lipopolysaccharide. The lateral flow detection device contains an area where an immobilized monoclonal antibody captures the Shiga toxin producing *E. coli* O157, leading to the formation of antibody-*E.coli* O157-antibody/gold conjugate (Figure 1.2). The test also contains an internal positive control.

**Figure 1.1 Eclipse™ device**

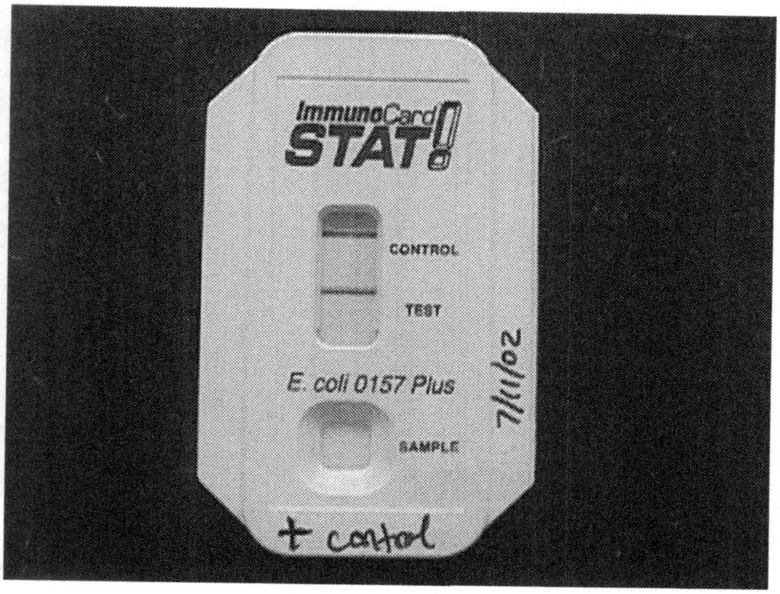


**Figure 1.2 Immunocard® Stat! device**

***Reveal® for E. coli O157:H7 Test (Neogen Corporation, Lansing, MI)***

The Reveal device uses the same lateral diffusion concept as described earlier. Capillary flow of the sample leads to capture of *E. coli* antigens by high specificity antibodies that have been conjugated to colloidal gold, thus leading to the formation of an antigen-antibody-chromogen complex. Migration of this complex across a lateral flow membrane and subsequent capture of the complex at a point further along the membrane, by using additional immobilized antibodies, generates a visible line indicating a positive reaction (Figure 1.3). The test also contains an internal positive control, which results in generation of a second line to validate that the test has functioned appropriately. This test has been used by the U.S. Department of Agriculture's Food Safety Inspection Service (FSIS) since October 1994 and over 39,000 samples have been analyzed with < 1% false positives.

***E. coli O157 Antigen Detection Test (Mionix, Naperville, IL)***

This kit is not actively manufactured, however expired kits were donated for evaluation in this study. This device combined an antibody-dye conjugate and solid phase antibodies to selectively identify *E. coli* O157. As the test sample flows through the device, the labeled antibody-dye conjugate binds to *E. coli* O157 antigen, forming a complex that binds to a stationary *E. coli* antibody located in the test zone. This produces a dark pink to purple band indicating a positive *E. coli* O157 test (Figure 1.4). In samples that do not contain *E. coli* O157:H7, no line is generated in the test zone and unbound conjugate binds to reagent in the control zone. The results are interpreted as negative when a band appears only in the 'control' window. According to the manufacturer, this assay has a sensitivity and specificity of >99% and is capable of detecting 0.08 CFU of *E. coli* O157 per gram of meat.

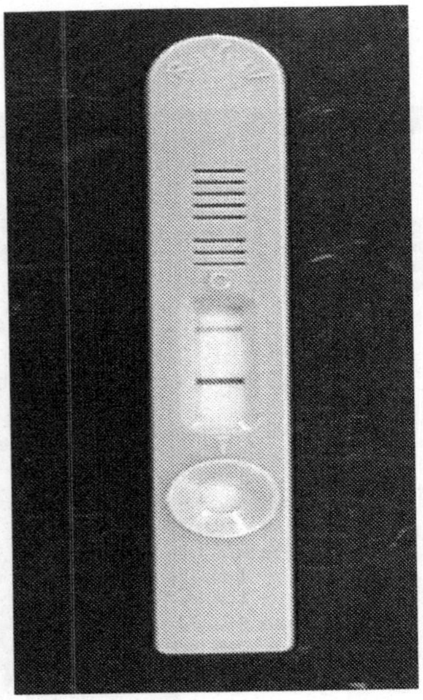

**Figure 1.3 Reveal® device**

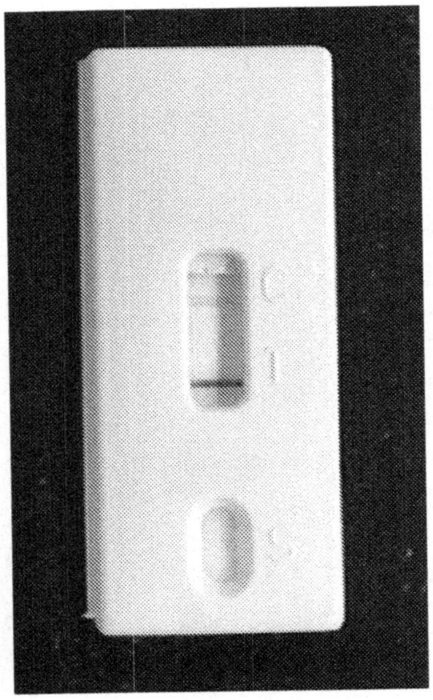

**Figure 1.4 Mionix device**

***BAX® E. coli O157:H7 assay (DuPont Qualicon, Wilmington, DE)***

The *E. coli* O157:H7 BAX® target sequences have been derived through random amplified polymorphic DNA (RAPD) analysis and the function of the target gene is not completely known. Sequence analysis suggests the amplified DNA fragment shows strong similarities to two related bacteriophage sequences. The first is from the lambdoid phage HK97, but no function is associated with the sequence. The second is a prophage *rus* sequence found in *E. coli* that is associated with production of a nuclease involved in recombination. The primer sequence has been reported to be specific for *E. coli* O157:H7 by screening 113 target and 176 non-target organisms, producing a 530-bp amplicon in 99% of the target and 0% of the non-target isolates (Johnson, Brooke, and Fritschel 1998). The BAX- *E. coli* O157:H7 assay has been "Performance Tested" Under the Association of Official Analytical Chemists Research Institute evaluation procedures. The assay apparently detected organisms in 96.5% of spiked ground beef samples, and out performed the best cultural method (39%) and immunodiffusion method (71.5%) (Johnson, Brooke, and Fritschel 1998).

The BAX ® assays are PCR-based screening tests for the detection of pathogenic microorganisms and utilize primers, as well as other PCR reagents, in tablet form, pre-packaged in PCR tubes. This standardized system employs simple sample preparation and minimizes liquid transfer to produce a fast, accurate, easy-to-use test. The BAX® assay runs on the Automated BAX® instrument for PCR amplification and detection (Figure 1.5). The computer-controlled instrument is a combination thermal cycler and fluorometer. The Automated BAX® System utilizes a closed-tube, fluorescent melting curve analysis for the detection and identification of specific PCR products (Tice *et al.*, 2000). This eliminates the need for gel electrophoresis, saves time and labor, and reduces the possibility of carryover contamination.

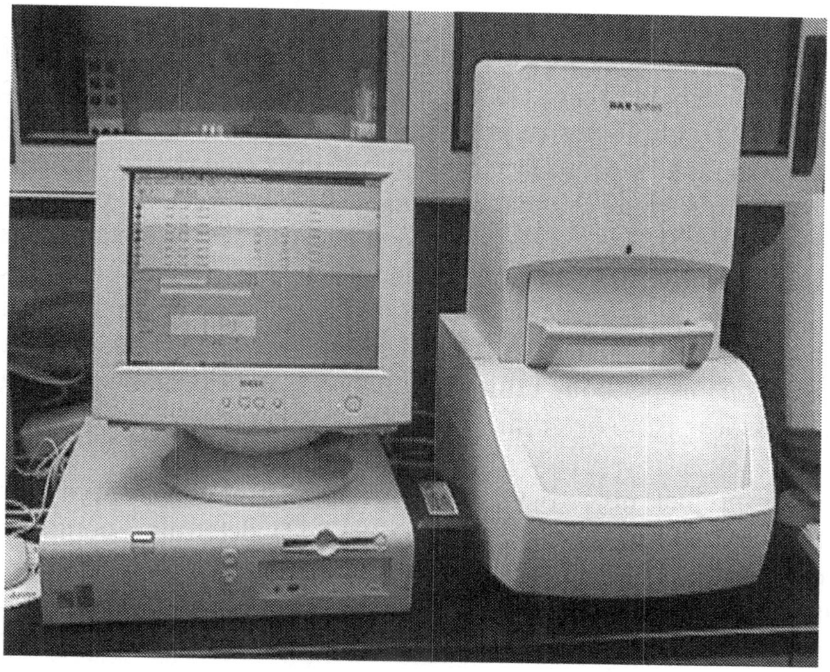

**Figure 1.5 The BAX® System**

SYBR Green, a double stranded DNA (dsDNA) intercalating dye, is incorporated into the PCR tablet and is used for the fluorescent melting curve detection of specific PCR products. The basis for melting curve analysis is that SYBR Green binds to dsDNA PCR products and fluoresces when exposed to an excitation light. The fluorescence within the PCR sample is measured while the temperature is slowly increased from 70°C to 95°C. The PCR product melt into two chains of single stranded DNA at a temperature that is characteristic of a specific PCR product. Once the dsDNA has melted, the SYBR Green dissociates from the DNA, leading to a significant reduction in fluorescence. The Automated BAX® System uses software to monitor the fluorescence, mathematically analyze the data and recognize and identify the temperature where a reduction in fluorescence occurs. This reduction in fluorescence is graphically displayed as a peak at the specific temperature where the fluorescence decreased. The software thereby determines if each sample is positive for the target organism (i.e., *E. coli* O157:H7) based upon the presence of a melting curve peak at the specific target temperature. If the sample is negative, then there is no yield of the PCR, and, as a result, there is no melting curve peak at the specified target temperature. The BAX® reaction tablets also contain a second set of primers and a small amount of a known DNA fragment, which serves as an internal positive control (INPC) (Farnum *et al.*, 2000). The INPC PCR product is generated whether the sample is positive or negative and has a melting temperature that is different from *E. coli* O157:H7 PCR product. The INPC provides a control reaction to demonstrate that the PCR cycling conditions were correct and that there was no inhibition of the PCR by sample matrices.

## EXPERIMENTAL PLAN AND OBJECTIVES

Based on the methods currently employed in water utility laboratories for compliance monitoring for the Total Coliform Rule, several different approaches were selected in this study for determining enrichment and isolation for recovery of *E. coli* O157:H7 from source and finished water concentrates. Membrane filtration and centrifugation were evaluated for concentrating large sample volumes. Lauryl tryptose broth, brilliant green bile broth, and

Colilert® media were examined as enrichment media for *E. coli* O157:H7. Various other media, including some specifically designed for *E. coli* O157:H7 enrichment, were also evaluated. Since *E. coli* O157:H7 would likely be present in only low numbers in environmental samples, the use of immunomagnetic separation (IMS) to aid in organism isolation was also assessed. Two types of IMS beads were evaluated, Dynabeads® anti-*E. coli* O157 (Dynal Inc., Lake Success, NY) and *E. coli* O157 A-Beads™ (Aureon Biosciences, Vienna, Austria). Both consist of paramagnetic beads with affinity purified antibodies against *E. coli* O157 covalently bound to their surface. After IMS, the samples were then subjected to the desired detection procedure for *E. coli* O157:H7. The overall method developed in one laboratory underwent a multi-laboratory round robin evaluation, using both utility laboratories and specialized research facilities, to identify procedural issues and to determine performance specifications and robustness.

The specific project objectives were to 1) determine sensitivity and specificity of proposed detection procedures for *E. coli* O157:H7, 2) determine compatibility of selected detection procedures with proposed isolation procedures for recovery of *E. coli* O157:H7 from source and finished water samples, and 3) evaluate the overall procedures in multiple utility laboratories to identify method(s) sensitivity and robustness.

# CHAPTER 2
# MATERIALS AND METHODS

## DESCRIPTION OF *E. COLI* O157:H7 ISOLATES

Four different isolates of *E. coli* O157:H7 were purchased from American Type Culture Collection (ATCC, Manassas, VA). The characteristics of these different isolates are summarized in Table 2.1. The majority of experimental work performed during the project used *E. coli* O157:H7 strain ATCC 35150. Cultures of *E. coli* O157:H7 were propagated using Tryptic Soy Broth (TSB) and any necessary dilutions were prepared using Phosphate Buffered Saline (PBS). Tryptic Soy Agar (TSA) plates were used for routine enumerations with the spread plate method according to the 20th edition of Standard Methods of the Examination of Water and Wastewater (APHA *et al.,* 1998).

**Table 2.1**
***E. coli* O157:H7 isolates used during this study**

| ATCC Number | Original Source | Comments |
|---|---|---|
| 35150 | Human feces | Causes hemorrhagic colitis |
| 43888 | Human feces | Lacks genes for Shiga toxins 1 or 2 |
| 43889 | Human feces | Causes hemolytic uremic syndrome. Produces Shiga toxin 2 |
| 43890 | Human feces | Produces Shiga toxin 1 |

## DESCRIPTION OF NON-TARGET ORGANISMS

To examine cross-reactivity of various detection methods, a total of five non-target organisms were chosen for use in the study, and the characteristics of these non-O157 organisms are listed in Table 2.2. These organisms were used in cross-reactivity and competition studies during evaluation of enrichment media, IMS and selected detection procedures, as well as during evaluation of the overall method. In addition, two selective media were also evaluated using these non-O157 organisms.

**Table 2.2**
**Non-target organisms used during this study**

| Organism | ATCC Number | Comments |
|---|---|---|
| *Salmonella choleraesuis* | 13076 | Type strain of *Salmonella enteritidis* |
| *Escherichia coli O25:K98:NM* | 43886 | Isolated in shipboard outbreak |
| *Escherichia coli* | 11229 | QC strain |
| *Enterobacter cloacae* | 13047 | QC strain for Sensititre, API and BBL products |
| *Citrobacter freundii* | 8090 | Produces restriction endonuclease *cfra*1. QC strain for Sensititre products |

## SENSITIVITY MEASUREMENTS OF COMMERCIAL DETECTION PROCEDURES

### Immunological Detection Procedures

#### *Eclipse™ E. coli O157:H7 Test (Eichrom Technologies, Inc, Darien, IL)*

The sensitivity of the Eclipse™ *E. coli O157:H7* test was determined using a single *E. coli O157:H7* strain (ATCC 35150) and spike doses of either $1x10^6$ or $1x10^5$ CFU per test kit. This procedure required boiling of the samples (10 min) prior to analyses. After boiling, the tubes were allowed to cool by placing them in a refrigerator briefly. Each test sample was handled carefully to avoid resuspension of settled debris and 100 μL of the supernatant (containing the target number of organisms) was placed into the sample well of the test device. Test results were visualized 20 min after addition of the sample to the sample well. The test was considered positive for the presence of *E. coli* O157:H7 when bands were visible in the test zone and control zone. In contrast, test devices containing only a band in the control zone were determined to be negative for the presence of *E. coli* O157:H7. The complete absence of a band in the control zone rendered the test invalid.

#### *ImmunoCard® Stat! E. coli O157:H7 Test (Meridian Diagnostics Inc., Cincinnati, OH)*

The sensitivity of the ImmunoCard® STAT! test was determined using two *E. coli* O157:H7 strains (ATCC 35150 and ATCC 43890), according to the manufacturer's instructions. This test kit was examined with spike doses of $1x10^6$ CFU and $1x10^5$ CFU using both strains of *E. coli* O157:H7. Unlike the other lateral flow assays, this test did not require boiling of the sample before use. Briefly, 100 μL of sample containing a known concentration of *E. coli* O157:H7 was added to the sample port of each cassette. The cassettes were then incubated for 10 min at room temperature. Samples were determined to be positive for *E. coli* O157:H7 based on the presence of a red/purple line in the test zone and a second line in the control zone. In contrast, absence of a red/purple line in the test zone (with presence of a line in the control zone) indicated the inability of the kit to detect *E. coli* O157:H7 at that given concentration. The complete absence of a band in the control zone rendered the test invalid.

#### *Reveal® for E. coli O157:H7 Test (Neogen Corporation, Lansing, MI)*

The sensitivity of the Reveal® *E. coli* O157:H7 test was determined using two *E. coli* O157:H7 strains (ATCC 35150 and ATCC 43890). While one isolate (ATCC 35150) was examined at three different spike doses ($1x10^5$, $1x10^4$, and $1x10^3$ CFU), the second isolate (ATCC 43890) was examined at two spike doses ($1x10^6$ and $1x10^5$ CFU). Briefly, the samples were placed in microfuge tubes and boiled in a water bath for 10 min prior to the assay. After boiling, the tubes were allowed to cool by placing them briefly in a refrigerator. Each test sample was handled carefully to avoid resuspension of settled debris and 100 μL of the supernatant (containing the target number of organisms) was placed into the sample well of the test device. Test results were visualized 15-20 min after addition of the sample to the sample well. The test was considered positive for the presence of *E. coli* O157:H7 when bands were visible in the test zone and control zone. In contrast, test devices containing only a band in the control zone were determined to be negative for the presence of *E. coli* O157:H7. The complete absence of a band in the control zone rendered the test invalid. In subsequent studies involving an evaluation of the overall method, 100 µL portions of enriched samples were boiled and loaded onto the sample port as described. In some cases, a small portion (less than 5%) of the bead-bacteria complex was analyzed by spread plate analysis before the boiling step. The samples were enumerated on TSA plates after incubation at 37°C for 16-20 h.

### *E. coli O157 Antigen Detection Test (Mionix, Naperville, IL)*

Although Mionix is no longer manufacturing this kit, they did donate expired tests (six months past the expiration date) for use in this study and their sensitivity was also examined. These kits were tested using a single strain of *E. coli* O157:H7 (ATCC 35150) and spike doses of either $1x10^5$ or $1x10^4$ CFU per test. This procedure required boiling of the *E. coli* samples (10 min) prior to analyses. After boiling, the tubes were allowed to cool by briefly placing them in a refrigerator. Each test sample was handled carefully to avoid resuspension of settled debris and 100 µL of the supernatant (containing the target number of organisms) was placed into the sample well of the test device. Test results were visualized 8-10 min after addition of the sample to the sample well. The test was considered positive for the presence of *E. coli* O157:H7 when a pink/purple band was visible in the test zone and a second pink/purple band was also observed in the control zone. In contrast, test devices indicating only a pink/purple band in the control zone were determined to be negative for the presence of *E. coli* O157:H7. The complete absence of a band in the control zone rendered the test invalid.

## Molecular Detection Procedure

### *BAX® E. coli O157:H7 Assay (DuPont Qualicon, Wilmington, DE)*

The sensitivity of the BAX® System was determined using a single strain of *E. coli* O157:H7 (ATCC 35150) according to the manufacturers instructions. Briefly, lysis reagent was prepared by preparing 1/80 dilution of the protease enzyme in the lysis buffer (both provided in the BAX® System kit). For example, using 12.5 µL of protease enzyme, the enzyme volume was adjusted to 1 mL with lysis buffer to obtain the desired working dilution in the lysis reagent. Lysis reagent (200 µL) was added to each test sample (typical volume 2-5 µL) and samples were incubated at 37°C (20 min) followed by 95°C (10 min). Samples were allowed to cool to room temperature and 50 µL of each sample lysate was transferred to individual PCR tubes containing BAX® PCR tablets. Following hydration of the tablets, the PCR tube strips were loaded into the BAX® cycler/detector. The BAX® System application contains preprogrammed cycling and detection conditions which are initiated by choosing the appropriate target pathogen (*E. coli* O157:H7) within the BAX® System application. Following completion of the cycling and detection program, data output was recorded qualitatively as either positive or negative for the presence of *E. coli* O157:H7 target DNA sequence. In some cases, a small portion (less than 5%) of the bead-bacteria complex was analyzed by spread plate analysis before the lysis procedure in order to quantitatively determine the inoculum size of organisms that were subjected to BAX® analysis. The spread plate samples were enumerated on TSA plates after incubation at 37°C for 16-20 h.

## EVALUATION OF GROWTH MEDIA AND GROWTH DYNAMICS OF *E. COLI* O157:H7

### Comparison of Commercially Available Media

A number of different media were examined to determine an effective growth enrichment media for *E. coli* O157:H7 (ATCC 35150). The media included Tryptic Soy Broth (TSB) (Becton, Dickinson and Co., Sparks, MD), BAX® Enrichment (Dupont Qualicon), Reveal® Enrichment (Neogen), Mēdi8™ (Eichrom Technologies) and mEC (Becton, Dickinson and Co.). All were prepared according to manufacturers instructions. Based on recommendations from the manufacturers, an additional comparison of TSB, mEC, BAX® Enrichment and Reveal® Enrichment supplemented with 1.5% Beef Extract was also examined. Media were prepared according to the manufacturers instructions and the Beef Extract was added to yield a final concentration of 1.5%. For each media type, approximately 100 mL volumes were aseptically transferred into 250 mL Erlenmeyer flasks and inoculated with either 10 or 100 CFU of *E. coli* O157:H7 (ATCC 35150). Flasks were prepared in duplicate – one for a growth curve at 37°C and one for a growth curve at 42°C. After inoculation, flasks were placed in an incubator set at the appropriate temperature, with shaking. For each media and temperature condition, sub-samples (5 mL) were removed at various time intervals (between 2 - 24 h) and organism number determined using the spread plate procedure on TSA plates. Where necessary, serial dilutions of the sub-samples were analyzed to ensure the number of colonies per plate did not exceed 300. The plates were incubated at 37°C overnight and enumerated the following day.

Lauryl Tryptose Broth (LTB), Brilliant Green Bile Broth (BGBB) and Colilert®, which are media used in total and fecal coliform analysis, were also evaluated for *E. coli* O157:H7 (ATCC 35150) propagation. LTB and BGBB were purchased from Becton Dickinson and were prepared according to Method 9221B (APHA *et al.,* 1998). Known volumes of the media (10 mL) were spiked with 2, 10 or 100 CFU and incubated at 37°C for 24 h. Serial dilutions of the cultures were made and 100 µL aliquots enumerated on TSA plates. Bottles of Colilert® (IDEXX, Westbrook, ME) media were prepared according to manufacturers instructions, inoculated with spike doses ranging from 1 to 40,000 CFU of *E. coli* O157:H7 (ATCC 35150), and grown for 24 h at 35°C. Spread plate analysis of 100 µL sub-samples, in duplicate, were used to enumerate the cultures. The plates were incubated at 37°C overnight and enumerated the following day.

### Enrichment of *E. coli* O157:H7 in Leighton Tubes

*E. coli* O157:H7 growth dynamics were evaluated in 125 mL Erlenmeyer flasks and Leighton tubes (Bellco Glass, Inc., Vineland, NJ). Stock volumes (typically 100 mL) of TSB were spiked with two different concentrations (1 CFU per mL or 0.1 CFU per mL) of *E. coli* O157:H7 (ATCC 35150). Sub-samples of the spiked stock (10 mL) were added to Leighton tubes (n=5) and 125 mL flasks (n=5) and incubated with end over end rotation (Leighton tubes) or shaking (flasks) for 3 h at 42°C. Dynabeads® anti-*E. coli* O157 (Dynal, Inc., Lake Success, NY) (100 µL) were added to each culture and the immunomagnetic separation (IMS) procedure (described later) followed at room temperature. Post-IMS, 5-10% of the bead-bacteria concentrate was analyzed on TSA plates by the spread plate procedure for samples containing an initial spike dose of 1 CFU per mL, whereas 25% of the concentrate was analyzed for samples containing an initial spike dose of 0.1 CFU per mL. The plates were incubated at 37°C overnight and enumerated the following day.

### Evaluation of Selective Media

In this study, enumeration of *E. coli* O157:H7 was usually performed on TSA plates, however the usefulness of two selective media was also examined. These two selective media were sMAC and CHROMagar™ O157 medium (Becton Dickinson, Cockeysville, MD). The sMAC medium utilizes the inability of *E. coli* O157:H7 to ferment sorbitol such that O157 colonies can be selected based on their morphology. *E. coli* O157:H7 colonies appear white, whereas other colonies display a magenta color. CHROMagar™ O157 medium was purchased as prepared plates and utilizes a combination of *E. coli* O157:H7's inability to ferment sorbitol fermentation and its lack of beta-glucuronidase activity for identification purposes. Chromogenic substrates within the medium allow for identification of O157 strains by their mauve colony formation. Other bacteria are either inhibited entirely or appear as colorless or blue-green colonies. All isolates (described in Table 2.1 and Table 2.2) were examined on the selective media. Plates were incubated according to manufacturer instructions and enumerated the following day.

## IMMUNOMAGNETIC SEPARATION (IMS) OF *E. COLI* O157:H7

### Effect of Sample Size, Bead Volume, Diluent and Incubation Temperature on IMS

The performance of two commercially available IMS kits [Dynabeads® anti-*E. coli* O157 (Dynal, Inc.) and *E. coli* O157 A-Beads ™ (Aureon Biosciences, Vienna, Austria)] were examined in this study. According to the manufacturer instructions, 20 µL of beads were added to 1 mL sample volumes spiked with $1x10^4$ CFU of *E. coli* O157:H7 (ATCC 35150), incubated for 10 min at room temperature, and the bead-bacteria complexes were recovered using magnets provided by the manufacturers. Recovery efficiencies were also evaluated using larger volumes (10 mL of PBS or TSB) of spiked samples and using increased volumes (100 µL or 200 µL) of beads. In these latter tests, bead incubation times were increased from 10 min to 1 h. The effect of an elevated incubation temperature (42°C) was also assessed. The samples were separated according to each manufacturer's recommended protocol. Following IMS capture and separation, each bead-organism conjugate was resuspended in 10 mL of PBS and 100 µL sub-samples were spread plated in duplicate using TSA plates. The plates were incubated at 37°C overnight and enumerated the following day. Controls for measuring IMS recovery efficiency were handled exactly as the sample processed by IMS, except that these controls did not contain IMS beads and were not subjected to magnetic separation. Comparison between the IMS samples and IMS controls enabled determination of IMS recovery efficiency.

### Matrix Effects on IMS Recoveries

To examine the impact of sample matrix on IMS recovery efficiency, 200 mL of autoclaved river water from four sites as well as three ground water samples were filtered through 0.45 µm MicroFunnel™ (Pall Gelman, Ann Arbor, MI) filters. The filters were eluted with 10 mL TSB and each sample eluant was spiked with $1x10^4$ CFU of *E. coli* O157:H7 (ATCC 35150). The addition of 100 µl of Dynabeads® anti-*E. coli* O157 immunomagnetic beads to each spiked sample was followed by incubation at 42°C for 1 h. Parallel control samples, consisting of 10 mL volumes of TSB, spiked with 1 x $10^4$ CFU of *E. coli* O157:H7 (ATCC 35150) and 100 μL of beads were incubated for 1 h at 42°C. All test samples and controls were magnetically separated, and the bead-bacteria complexes resuspended in 10 mL PBS. For each concentrate, triplicate 100 μL aliquots were plated onto individual TSA plates. The plates were incubated at 37°C overnight and enumerated the following day. Recovery efficiencies were calculated by comparing river and ground water sample enumerations to control sample enumerations.

**Effect of Non-Target Organisms on IMS Recovery**

The specificity of the IMS step was assessed by performing recovery trials using *E. coli* O157:H7 and various doses of non-O157 organisms in 10 mL volumes of 1X TSB (i.e., 1x$10^4$ CFU of *E. coli* O157:H7 (ATCC 35150), 1x$10^4$ CFU of a mixture of non-O157 organisms or a combination of *E. coli* O157:H7/non-O157 organisms). The non-O157 organism mixture consisted of *E. coli* ATCC 11229, *E. coli* O25:K98:NM, *Salmonella choleraesuis*, *Enterobacter cloacae* and *Citrobacter freundii*. Following IMS with Dynabeads®, the bead-bacteria complex was re-suspended in 10 mL of PBS and 100 μL aliquots enumerated in duplicate using the spread plate method on TSB, sMAC and CHROMagar™ plates. The non-specific capture of competing organisms was determined by comparing colonies (based on morphology) in samples containing all organisms to samples containing only *E. coli* O157:H7.

**Evaluating Simultaneous Enrichment and Capture of *E. coli* O157:H7**

Five 10 mL sub-samples of TSB, spiked with an initial concentration of either 1 CFU per mL or 0.5 CFU per mL of *E. coli* O157:H7 (ATCC 35150), were dispensed into individual Leighton tubes. Comparisons between a separate enrichment and capture step with a simultaneous enrichment/capture step were performed. For the former, samples were mixed on a Dynal sample mixer for either 3 h or 6 h at 42°C, whereas, for the latter, each sample was mixed on a Dynal sample mixer for either 2 h or 5 h at 42°C. Dynabeads® anti-*E. coli* O157:H7 (100 μL) were added to each tube and incubation (42°C) with mixing was continued for an additional 1 h. Each sample was magnetically separated, the bead-bacteria complex resuspended in sterile PBS and duplicate 100 μL sub-samples of each concentrate (appropriately diluted where necessary) plated onto TSA plates. The plates were incubated at 37°C overnight and enumerated the following day.

## PRELIMINARY TRIALS TO DETERMINE METHOD SENSITIVITY

TSB sub-samples (10 mL) were spiked with an initial concentration of either 1.0 CFU per mL, 0.5 CFU per mL or 0.25 CFU per mL of *E. coli* O157:H7, and dispensed into individual Leighton tubes. Each sample was mixed on a Dynal sample mixer for either 2 h or 5 h at 42°C. Dynabeads® anti-*E. coli* O157:H7 beads (100 μL) were added to each tube and incubation (42°C) with mixing was continued for an additional 1 h. Each sample was subjected to IMS, the bead-bacteria complex resuspended in sterile PBS and appropriately diluted sub-samples of each concentrate plated onto TSA plates. The plates were incubated at 37°C overnight and enumerated the following day.

## EVALUATION OF CONCENTRATION METHODS

It was anticipated that the overall method for rapid detection of *E. coli* O157:H7 would likely begin with a water sample ranging in volume from 50 mL to 1 L. It would not be feasible to use such a large volume with other components of the overall method, specifically the propagation-IMS recovery step, which requires concentration of the water sample to a suitable working volume. Two methods typically used for sample concentration that were evaluated in this study were membrane filtration and centrifugation.

### Centrifugation

Centrifugation studies were performed using 50 mL volumes of either Milli-Q Ultrapure (MQ) water (Millipore, Bedford, MA) or 0.2X TSB spiked with either $1x10^4$ CFU or 100 CFU of *E. coli* O157:H7 (ATCC 35150). These samples were centrifuged (15 min at 3000 x g) in 225 mL conical centrifuge tubes (Falcon, Becton Dickinson). Samples were aspirated down to 10 mL volumes, vortexed and 100 µL volumes analyzed by the spread plate procedure. Recoveries were determined relative to control 10 mL volumes, which received the same spike dose but were not processed by centrifugation. Additional trials analyzed 200 mL volumes of MQ, 1X TSB, or autoclaved source water from three sites (with or without various concentrations of TSB), each spiked with 1 x $10^4$ CFU *E. coli* O157:H7. The three different river samples were autoclaved (20 min at 121°C and 16 psi) to enable differentiation of the laboratory spike dose from naturally occurring *E. coli* O157:H7. All samples were centrifuged in Falcon 225 mL conical centrifuge tubes (15 min at 3000 x g), aspirated down to 10 mL volumes, the pellets resuspended by vortexing and 100 µL volumes analyzed by the spread plate method. In addition, one trial evaluated the effect of homogenization of the pellet on the recovery efficiency. Using a Polytron PT1200, the samples were homogenized at approximately 7,500 rpm for 30 sec. Aliquots were plated before and after homogenization.

### Membrane Filtration (MF)

MQ water or autoclaved river water (200 mL volumes) from up to three sites was individually spiked with $1x10^4$ CFU of *E. coli* O157:H7 (ATCC 35150). The three different river samples were autoclaved (20 min at 121°C and 16 psi) prior to use to differentiate the laboratory spike dose from naturally occurring *E. coli* O157:H7. The spiked samples were filtered through 0.45 µm mixed cellulose ester filters, MicroFunnel™ (Pall Gelman, Ann Arbor, MI). Each filter was eluted with 10 mL of TSB and 100 µL of each filter concentrate was plated onto TSA plates in duplicate.

#### *Evaluating MF Techniques*

To identify potential areas for organism loss during membrane filtration, three methods for filtering spiked samples, two different lots of filters, and two different elution time intervals were examined. Low spike doses (8-20 CFU) were filtered through 0.45 µm MicroFunnel™ filters. Three types of filters were generated. One filter was simply pre-wet with MQ water and the *E. coli* O157:H7 (ATCC 35150) applied directly onto the filter, the filter removed from the housing, and transferred to a 50 mL conical tube. For the second filtration technique, direct application of the spike dose onto the filter was followed by filtration of 200 mL of MQ water through each filter. For the third filtration technique, 200 mL of MQ water containing the spike dose was filtered through each filter in the usual manner. All filters were eluted using 10 mL of 1X TSB in a 50 mL conical tube. Elution was performed for either 10 min or 2 h at 42°C and the samples transferred to Leighton tubes. The samples eluted for 10 min were incubated for 2 h at 42°C (to mimic the usual overall method procedure). Dynabeads® anti-*E. coli* O157 (100 μL) were added to each sample following the 2 h incubation (either in the Leighton tube or 50 mL conical). Samples were transferred to Leighton tubes and incubated at 42°C for an additional 1 h. The samples were separated according to manufacturer's instructions, the bead-bacteria complexes resuspended in 1 mL of PBS, and 100 μL aliquots plated onto duplicate TSA plates. Plates were incubated at 37°C overnight and enumerated the following day Recovery efficiencies for each matrix were calculated relative to their growth controls. Growth controls consisted of 10 mL of 1X TSB in Leighton tubes, spiked with an equal number of *E. coli* O157:H7 and incubated for 2 h at 42°C, followed by addition of beads and further incubation for 1 h at 42°C. Following IMS, the bead-bacteria complexes were resuspended in 1 mL of PBS and 100 μL aliquots were plated onto duplicate TSA plates.

**Concentration, Propagation and IMS**

MQ water, finished water containing 0.1% sodium thiosulfate (to neutralized residual chlorine) and autoclaved ground and river water samples (200 mL volumes) were each spiked with low numbers (8-20 CFU) of *E. coli* O157:H7 (ATCC 35150) and concentrated by either centrifugation or membrane filtration. Samples subjected to centrifugation were prepared in Falcon 225 mL conical centrifuge tubes (Becton Dickinson) and received 20 mL of 10X TSB. These suspensions were centrifuged for 15 min at 3000 x g. Samples were aspirated down to 1 mL volumes, 9 mL of 1X TSB added and the pellets resuspended by vortexing. Samples subjected to membrane filtration were prepared in sterile glass bottles before filtering through a 0.45 μm MicroFunnel™. Each filter was eluted with 10 mL of TSB. The resuspended sample concentrates were transferred to individual Leighton tubes and incubated for 2 h at 42°C. Dynal anti-*E. coli* O157 immunomagnetic beads (100 μL) were added to each sample concentrate and incubated at 42°C for an additional 1 h. For centrifuged samples, growth controls consisted of 9 mL TSB, 0.1 mL 10X TSB and 0.9 mL MQ water. For membrane filtration, growth controls consisted of 10 mL of 1X TSB. All growth controls were spiked with 10 CFU and incubated for 2 h at 42°C, followed by addition of beads and incubation for 1 h at 42°C. Following IMS, the bead-bacteria complexes were re-suspended in 1 mL of PBS and 100 μL aliquots were plated onto TSA plates in triplicate. Plates were incubated at 37°C overnight and enumerated the following day. Recovery efficiencies were calculated by comparing the enumeration results of each matrix to that of its corresponding growth control.

**Effect of Non-Target Organisms on Overall Enrichment and IMS**

Aliquots of MQ water (200 mL) were spiked with low numbers (8-20 CFU) of *E. coli* O157:H7 (ATCC 35150). Some samples also received an additional spike of 100 CFU, 1,000 CFU or 10,000 CFU of a competing non-O157 organism mixture (*E. coli* ATCC 11229, *E. coli* O25:K98:NM, *Salmonella choleraesuis*, *Enterobacter cloacae* and *Citrobacter freundii*). The samples were concentrated by MF and the filters eluted with 10 mL 1X TSB. The concentrates were incubated for 2 h at 42°C, followed by the addition of Dynabeads® (100 μL) and additional incubation (1 h at 42°C). Following IMS, the bead-bacteria complexes were resuspended in 1 mL of PBS and 100 μL aliquots were plated onto either sMAC plates or CHROMagar™ O157 plates in duplicate. Plates were incubated at 37°C overnight and enumerated the following day. The effect of competing non-O157 organisms on method sensitivity was evaluated by comparing the number of presumed *E. coli* O157:H7 colonies (based on morphology) in samples containing the mixture of non-O157 organisms to that of samples that contained only *E. coli* O157:H7. This was achieved by enumeration of only white colonies on sMAC plates and mauve colonies on CHROMagar™ plates.

## CONFIRMATION PROCEDURES

### Latex Agglutination

The *E. coli* O157 Latex Test (Oxoid Inc, Ogdensburg, NY) and the Murex Wellcolex *E. coli* O157:H7 (Remel Inc, Lenexa, KS) were evaluated as confirmatory tools for presumptive *E. coli* O157. The Oxoid test is designed to identify members of the *E. coli* serogroup O157 through the use of a specific rabbit antibody to the O157 somatic antigen. As such, this kit may identify *E. coli* O157 organisms other than *E. coli* O157:H7. The Murex Wellcolex test is designed to test for both the O157 lipopolysaccharide antigen and the H7 antigen though the use of two separate rabbit antibodies. Tests from both manufacturers required emulsification of a selected colony in a drop (50 µL) of saline on a test card provided in the kit. Drops of latex reagents were added to the colony suspension followed by agitation (rocking motion) of the card for 30 sec to 1 min. A positive reaction was noted by agglutination of the latex particles into discernible clumps, whereas a negative reaction resulted in a smooth appearing suspension. Each kit provided positive and negative controls as well as latex controls to help identify auto-agglutinating organisms that could lead to false-positive reactions when testing samples. Both tests were examined by using 50 µL suspensions of known numbers ($1 \times 10^3$ to $1 \times 10^8$ CFU) of *E. coli* O157:H7 (ATCC 35150).

## SINGLE LABORATORY VALIDATION OF OVERALL METHOD

### Detection Using the Reveal® Device

Individual 200 mL sample volumes of MQ water, raw water and tap water were evaluated by the Reveal® method. *E. coli* O157:H7 was added to each water sample at spike doses ranging from 0.05 CFU per mL to 5.0 CFU per mL. Each spiked sample was concentrated by membrane filtration and the membranes eluted by agitation in 10 mL TSB. The 10 mL sample concentrates were transferred to Leighton tubes and incubated at 42°C for 5 h. Dynabeads® anti-*E. coli* O157 (100 µL) were added to the samples and incubation continued for an additional 1 h. Magnetic separation was performed according to manufacturer instructions and the bead-bacteria complex resuspended in 100 µL of PBS. Each 100 µL concentrate of the bead-bacteria complexes was boiled for 10 min and, after cooling, the supernatant loaded onto the Reveal® devices. The test devices were allowed to develop for 10-15 min and results recorded. For a small sub-set of samples, a small portion (less than 5%) of the bead-bacteria complex was analyzed by spread plate analysis before the boiling step. The samples were enumerated on TSA plates following incubation at 37°C for 16-20 h.

## Detection Using the BAX® System

Individual 200 mL sample volumes of MQ water, raw water and tap water were evaluated by the BAX® method. *E. coli* O157:H7 was added to each water sample at spike doses ranging from 0.05 CFU per mL to 5.0 CFU per mL. Each spiked sample was concentrated by membrane filtration and the membranes eluted by agitation in 10 mL TSB. The 10 mL samples were transferred to Leighton tubes and incubated at 42°C for 2 h. Dynabeads® anti-*E. coli* O157 (100 μL) were added to the samples and incubation continued for an additional 1 h. Magnetic separation was performed according to manufacturer instructions and the bead-bacteria complex resuspended in 1 mL of PBS. Approximately 50% of the 1 mL was reserved for later analysis, such as spread plate enumeration. The remaining 50% was separated with the IMS magnet, the complex resuspended in 50 μL of BAX® lysis buffer and the lysis procedure initiated. The BAX® tablets were rehydrated using these 50 μL volumes of bead-bacteria lysate. Care was exercised in avoiding addition of immunomagnetic beads to the tablets. The BAX® program was initiated and results obtained upon completion of the thermocycling process.

## Impact of IMS Bead Carryover on BAX® Assay Sensitivity

BAX® lysis reactions were prepared using 50 μL of lysis reagent and a spike dose of 10 CFU, 25 CFU, 50 CFU or 100 CFU. Some reactions were processed through the lysis step without addition of Dynabeads® anti-*E. coli* O157 immunomagnetic beads, whereas in other reactions an immunomagnetic bead pellet equivalent to 20 μL (10 CFU only) or 50 μL was also added. Following the lysis step (20 min at 37°C and 10 min at 95°C), all lysed samples (including some containing 20 μL or 50 μL of beads) were added to individual BAX® tablets. The BAX® program was initiated and results obtained upon completion of thermocycling.

## Modification of the Method Incorporating Reveal® Detection

Discussion with utility labs before initiation of the multi-lab round robin indicated that it would be inconvenient to perform sample enrichment at 42°C, due to their need to maintain incubators at 35°C for compliance testing. This necessitated an evaluation of the sample enrichment step at an incubation temperature of 35°C to assess how it impacted the sensitivity of the overall method utilizing the Reveal® Detection system.

# MULTI-LABORATORY ROUND ROBIN

## Overall Method Using the Reveal® Detection Assay

A total of nine utility labs, located in CA, IL, MO, PA, NJ, WV and IA, were selected to process finished and river water samples using the Reveal® detection method for the multi-laboratory round robin. Each lab processed eight 200 mL volumes of each water matrix. Each individual sample was spiked with *E. coli* O157:H7 spike doses provided by the project team. At each round robin facility, the spiked samples were concentrated by membrane filtration and the membranes eluted with 10 mL TSB. The 10 mL samples were transferred to Leighton tubes and incubated at 35°C for 5 h. Dynabeads® anti-*E. coli* O157 (100 μL) were added to the samples and incubation continued for an additional 1 h. IMS was performed according to manufacturer instructions and the bead-bacteria complex resuspended in 100 μL of PBS. The complexes were boiled for 10 min and allowed to cool. The supernatant was then loaded onto the Reveal®

devices. The test devices were allowed to develop for 10-15 min and results recorded. Utility labs were also provided with positive and negative controls for the Reveal® devices.

**Overall Method Using the BAX® Detection Assay**

A total of four research labs, located in HI, NC, MI and MN, were chosen to process finished and river water samples by the BAX® detection method for the round robin. Each lab processed eight 200 mL volumes of each water matrix. Each individual sample was spiked with *E. coli* O157:H7 spike doses provided by the project team. At each round robin facility, the spiked samples were concentrated by membrane filtration and the membranes eluted with 10 mL TSB. The 10 mL samples were transferred to Leighton tubes and incubated at 42°C for 2 h. Dynabeads® anti-*E. coli* O157 (100 µL) were added to the samples and incubation continued for an additional 1 h. IMS was performed according to manufacturer instructions and the bead-bacteria complex resuspended in 1 mL of PBS. A 50% sub-sample of each 1 mL concentrate was reserved for later analysis or for shipment back to the project team. The remaining 50% was separated with the IMS magnet, the complex resuspended in 50 µL of BAX® lysis buffer and the lysis procedure conducted. Individual BAX® tablets were rehydrated using these 50 µL volumes of bead-bacteria lysate. The round robin labs were instructed in their standard operating procedures (provided by the research team) to exercise care during this step to minimize the likelihood of carrying over immunomagnetic beads with the sample into the BAX® tablets. The BAX® program was initiated and results obtained upon completion of the pre-defined thermocycling step. The labs were provided with positive and negative stock solutions that were used as controls in the BAX® reactions.

# CHAPTER 3
# RESULTS AND DISCUSSION

## SENSITIVITY MEASUREMENTS OF COMMERCIAL DETECTION PROCEDURES

### Immunological Detection Procedures

The sensitivity of four commercially available immunological detection kits, which use a lateral diffusion format procedure, was examined. A known number of *E. coli* O157:H7 organisms were prepared according to kit directions and administered in 100 μL volumes to the test devices. After 10-15 min incubation, the test devices were scored and the results presented in Table 3.1. These data indicated that the ImmunoCard® stat test kit required $1 \times 10^6$ CFU of either of the two strains of *E. coli* O157:H7 to generate a medium strength positive signal, whereas with the Eclipse™ test kits and one strain (ATCC 35150), a signal was generated with $1 \times 10^6$ CFU *E. coli* O157:H7. Both these kits failed to produce a signal when $1 \times 10^5$ CFU of *E. coli* O157:H7 were examined. In contrast, both the Mionix and Reveal® test kits yielded strong positive signals when a spike dose of $1 \times 10^5$ CFU per test of one *E. coli* O157:H7 isolate (ATCC 35150) were used. Although the Reveal® test continued to yield a weak signal with $1 \times 10^4$ CFU of the same *E. coli* O157:H7 isolate, two of three tests with the Mionix test kit produced no signal at the same organism concentration. In one of the three tests with the Mionix kits, no band was observed in the control zone, thus invalidating the test. The performance of this kit may have been impacted by the fact that it had exceeded its expiration date. Using the second *E. coli* O157:H7 strain (ATCC 43890), approximately $1 \times 10^6$ CFU was required to produce a signal of the same intensity as the first *E. coli* O157:H7 strain (ATCC 35150) for the Reveal® test kit. During these latter tests, a weaker control band signal than normal was also generated. Weaker control bands with the second *E. coli* O157:H7 isolate suggest it is likely that inferior kit performance was the cause of the poor signal strength rather than an actual difference in sensitivity to different *E. coli* O157:H7 isolates.

**Table 3.1**
**Sensitivity determination for lateral diffusion assays**

| | | Spike dose (CFU) per test | | | |
|---|---|---|---|---|---|
| **Test Kit** | ***E.coli* strain** | $1 \times 10^6$ | $1 \times 10^5$ | $1 \times 10^4$ | $1 \times 10^3$ |
| ImmunoCard® STAT! | 35150 | ++ | - | NA | NA |
| | 43890 | ++ | - | NA | NA |
| Mionix | 35150 | NA | +++ | -* | NA |
| Eclipse™ | 35150 | ++ | - | NA | NA |
| Reveal® | 35150 | NA | +++ | + | - |
| | 43890 | +++ | + | NA | NA |

**Key:**
- +++ Strong positive signal
- ++ Medium positive signal
- + Weak positive signal;
- - No signal
- NA Not Applicable (or not examined)
- * One of the replicates was invalid, determined by absence of a control band

### Molecular Detection Procedure

BAX® System sensitivity was examined using spike doses ranging from $1x10^5$ CFU to 0.1 CFU of *E. coli* O157:H7 (ATCC 35150) (Figure 3.1). Results indicated that spike doses of 10 CFU or higher yielded positive BAX® results 100% (n=14/14) of the time. Spike doses of 6 or 8 CFU yielded 80% and 65% positive rate respectively whereas spike doses between 2 - 5 CFU yielded 20% - 40% positives. Based on these data, it was evident that detection with the BAX® system necessitated a minimum of 10 CFU of *E. coli* O157:H7 per BAX® reaction to ensure 100% of the positive samples were successfully detected.

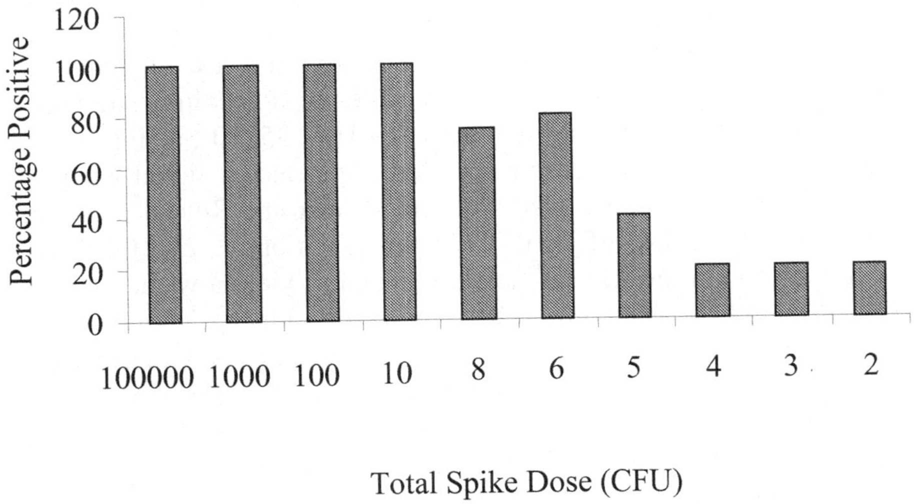


**Figure 3.1 *E. coli* O157:H7 sensitivity measurement with BAX® assay**

## CROSS REACTIVITY STUDIES

### Reveal® and BAX® Detection Procedures

Five non-O157 organisms were tested on the Reveal® detection device and by the BAX® assay to address potential cross-reactivity issues. *Citrobacter freundii*, *Enterobacter cloacae*, *Salmonella choleraesuis*, *E. coli* ATCC 11229 and *E. coli* O25:K98:NM were used at a spike dose of $1x10^6$ CFU and analyzed directly on the Reveal® and BAX® assays. Using the Reveal® assay, these five non-O157 organisms did not show a band in the test zone, indicating no cross-reactivity. During these trials, both positive and negative controls for each assay indicated appropriate results. Using the BAX® assay, reactions processed with each of the five non-O157 organisms yielded results negative for *E. coli* O157:H7, also indicating no cross-reactivity. While results were encouraging in that each detection procedure showed no cross-reactivity with the five non-target organisms, at very high spike doses, it should be emphasized that this study could not perform an exhaustive investigation of the myriad of organisms that could potentially be present in environmental water samples. As a caveat, it will be important to ensure appropriate controls are in place during employment of these detection procedures to ensure the signals being generated are from putative *E. coli* O157:H7 rather than cross reacting species of organisms.

### Impact of Non-target Organisms on Detection Procedure Sensitivity

The non-target organisms used for the cross reactivity studies were also used in combination with known numbers of *E. coli* O157:H7 to determine whether interference from the non-target organisms impacted the overall sensitivity of the detection procedure for the target *E. coli* O157:H7. A concentration of $2 \times 10^4$ CFU of *E. coli* O157:H7 and a preparation containing approximately $0.2 \times 10^6$ CFU of each of the five non-O157 organisms (i.e., total of $1 \times 10^6$ non-target organisms) were examined on Reveal® devices in duplicate. The Reveal® assay demonstrated a positive signal for the presence of *E. coli* O157:H7 and doubling the non-target organisms concentration (i.e., $2 \times 10^6$ CFU) per test device also did not demonstrate interference in the sensitivity of detecting *E. coli* O157:H7. In contrast, with the BAX® assay reactions containing a total of $1 \times 10^6$ non-target organisms and 10 CFU of *E. coli* O157:H7, sensitivity appeared to be affected by the presence of non-O157 organisms. The percentage of positive samples declined from 100% to 68% when 10 CFU of *E. coli* O157:H7 were spiked per reaction. However, when higher spike doses (›25 CFU) of *E. coli* O157:H7 were combined with $1 \times 10^6$ CFU of non-target organisms, the percentage positive with the BAX® system was 100%. The controls yielded the appropriate results in every trial. Using the 10 CFU spike doses of *E. coli* O157:H7, it is unclear whether failure to detect 100% of the samples as positive with the BAX® system was due to interference from the non-target organisms or a reflection of the difficulties associated with accurately delivering a low spike dose of of the target organisms into each BAX® reaction. Typically, the volume of suspension used for delivering the target dose of 10 CFU was 2-3 µL. Although spike dose verification by spread plating indicated a range of 9-11 CFU was delivered per reaction, it is possible that the actual aliquots delivered in the BAX® assay had greater variability.

## EVALUATION OF GROWTH MEDIA AND GROWTH DYNAMICS OF *E. COLI* O157:H7

### Comparison of Commercially Available Media

A number of different media were examined to determine the most effective growth enrichment media for *E. coli* O157:H7 (ATCC 35150) at 37°C and 42°C using two different initial spike doses (1 CFU per mL and 0.1 CFU per mL). Sub-samples taken at specific time intervals were enumerated on TSA spread plates. Initial studies compared TSB, BAX® enrichment media, Reveal® enrichment media, mEC, and Medi8. Analysis of cultures produced with these media showed propagation rates ranging from 120 CFU per mL with BAX® enrichment media to $5 \times 10^4$ CFU per mL with TSB after 6 hours of incubation at 42°C using an initial spike dose of 1 CFU per mL. Propagation of *E. coli* O157:H7 was greater at 42°C than 37°C in all media tested except mEC. Overall the propagation rates in these enrichment media have been ranked in Figure 3.2, ranging from the most effective propagation media to the least effective propagation media. TSB, which non-selectively promotes growth of various organisms, yielded considerably higher propagation rates of *E. coli* O157:H7 (up to 416 times greater) than the other selective media tested.

**Figure 3.2 Enrichment capabilities of growth media for *E. coli* O157:H7**

Manufacturers of the BAX® medium have previously noted an increased performance when the BAX® medium is supplemented with 1.5% Beef Extract. As a result, performance of some of the previously tested media was compared with and without 1.5% Beef Extract. After 6 h of incubation, supplementation with Beef Extract yielded approximately 30-fold increase in *E. coli* O157:H7 concentrations with the BAX® media (120 CFU per mL compared to 3500 CFU per mL with addition of 1.5% Beef Extract). Addition of Beef Extract to the Reveal® media produced a 2.5 fold increase in organisms (2450 per mL compared to 6200 per mL after 6 h). However, addition of Beef Extract had only a negligible effect on the other media tested. Despite improvement of the BAX® medium with Beef Extract, performance of this media and other media continued to be significantly poorer than TSB.

In addition to the media listed in Figure 3.2, three media typically used in fecal coliform analysis [Lauryl Tryptose Broth (LTB), Brilliant Green Bile Broth (BGBB) and Colilert®] were evaluated for enrichment of *E. coli* O157:H7. Using an initial spike dose of 1 CFU per mL, all three cultures were incubated for 24 h at 37°C. After propagation the cultures were enumerated by the spread plate method on TSA plates. The titers of the LTB, BGBB and Colilert® cultures were $3.1 \times 10^8$ CFU per mL, $1.1 \times 10^9$ CFU per mL and $2.5 \times 10^8$ CFU per mL, respectively. Propagation in these media compared favorably to TSB, which produced a concentration of $2.1 \times 10^9$ CFU per mL after 24 h incubation at 37°C. In addition, direct analysis of sub-samples of the *E. coli* O157:H7 amplified in Colilert® media, using the Reveal® lateral flow detection assay, indicated the presence of strong positive signals. Analysis of these same sub-samples by the BAX® assay also resulted in positive reactions. This indicated that the Colilert® media did not pose an inhibitory effect for either the Reveal® or BAX® detection procedures. *E. coli* O157:H7 enrichment data from these media that are typically used in fecal coliform analysis were considered encouraging. It provides the potential for water utilities performing routine compliance total coliform analysis to directly (or following further enrichment) use TC samples for determining the presence of *E. coli* O157:H7.

**Growth Dynamics of *E. coli* O157:H7 in TSB**

To optimize the sample processing time for the overall method, achieve the target method sensitivity and generate sufficient organisms to yield a 100% positive rate with the selected detection procedures, the growth dynamics of *E. coli* O157:H7 (ATCC 35150) in TSB were determined. The results have been presented in Figure 3.3 using an initial spike dose of 1 CFU per mL of TSB at both 37°C and 42°C.

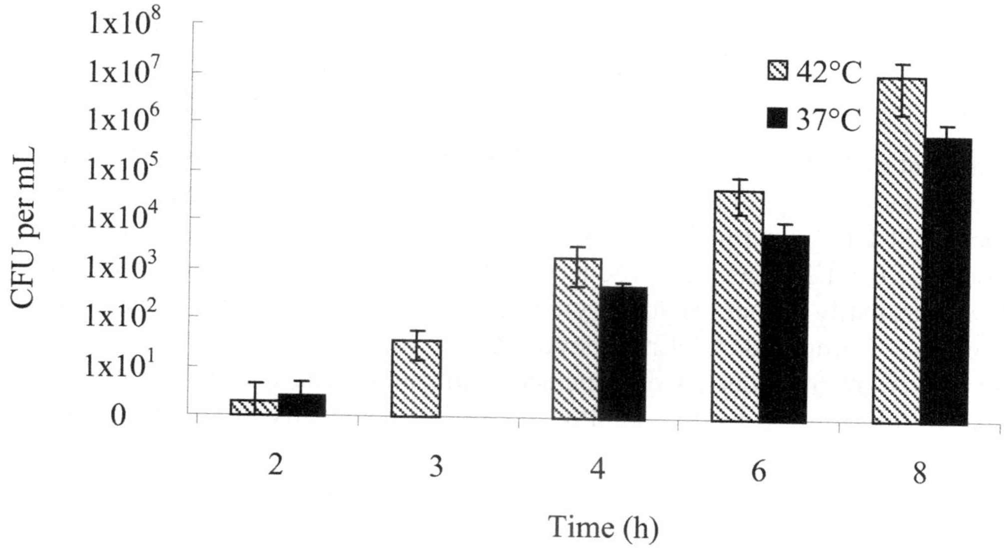


**Figure 3.3 Growth dynamics of *E. coli* O157:H7 in TSB**

After 3 h incubation at 42°C, the of *E. coli* O157:H7 concentrations reached approximately 100 CFU per mL and continued to demonstrate a consistent increase with concentrations approximating $0.5 \times 10^5$ CFU per mL after 6 h incubation and $1 \times 10^7$ CFU per mL after 8 h incubation. Growth dynamics at 37°C indicated similar propagation to that at 42°C following 2 h and 4 h incubation; however, between 4 h - 8 h incubation, the rate of propagation was markedly lower than that at 42°C. Using these growth dynamics data with an initial spike dose of 1 CFU per mL, it was estimated that a 3 h enrichment would yield sufficient quantities of *E. coli* O157:H7 (>100 CFU per mL) to generate a positive signal with the BAX® assay whereas a 6 h enrichment would be required to generate sufficient organisms ($>1 \times 10^4$ CFU per mL) for a positive signal with the Reveal® assay. The Reveal® detection procedure requires 30 min for completion, whereas the BAX® assay requires a minimum of 4 h. When accounting for sample manipulations and enrichment times of 3 h or 6 h as well as detection assay times of 4 h or 30 min, it was estimated that the total time required for the overall procedure would likely be 8 h - 10 h. As a result, further refinements to the procedure were necessary to ensure the overall procedure could be completed in single working day.

**Enrichment of *E. coli* O157:H7 in Leighton Tubes**

Initially relative performance of various growth media and determination of *E. coli* O157:H7 growth dynamics were performed in 125 mL Erlenmeyer flasks while being continuously mixed on a platform shaker. Subsequently enrichment of *E. coli* O157:H7 (ATCC 35150) in Leighton tubes was compared to Erlenmeyer flasks using initial spike doses of 0.1 CFU per mL and 1 CFU per mL in TSB at 42°C. The results are shown in Table 3.2.

**Table 3.2**
**Enrichment of *E. coli* O157:H7 in Leighton tubes versus flasks**

| Initial spike dose (CFU per mL) | Leighton tube (CFU per mL) | 125 mL flask (CFU per mL) |
|---|---|---|
| 1.0 | 174 ± 107.5 (n=20) | 149 ± 115.5 (n=20) |
| 0.1 | 15 ± 20.5 (n=10) | 10 ± 10 (n=10) |

Pre-enrichment of *E. coli* O157:H7 in Leighton tubes produced comparable results to pre-enrichment using flasks. Using an initial spike dose of 1.0 CFU per mL, the mean propagation in Leighton tubes (174 CFU per mL) was marginally higher than that of flasks (149 CFU per mL). A spike dose of 0.1 CFU per mL produced similar results, with propagation in Leighton tubes producing a marginally higher organism yield (15 CFU per mL) than propagation in flasks (10 CFU per mL). Compared to Erlenmeyer flasks, use of Leighton tubes would be advantageous in that transfer of enriched suspensions of pathogenic organism can be avoided. This is not only useful strategy from a health and safety perspective of handling an human pathogen, but also in minimizing unnecessary transfer of organisms thus helping to reduce potential loss of organisms and improving overall method sensitivity.

## IMMUNOMAGNETIC SEPARATION (IMS) OF *E. COLI* O157:H7

### Comparison of Sample Size, Bead Volume and Diluent on IMS

*E. coli* O157:H7 contaminated environmental water samples are likely to contain low concentrations (i.e. 0.04 CFU per mL to 0.2 CFU per mL) of organisms. Extrapolation from growth curves, using an initial spike dose of 1 CFU per mL, suggested that following propagation in TSB the final *E. coli* O157:H7 concentrations were unlikely to be sufficient to produce positive reactions by the selected detection methods. For example, using an initial concentration of 0.04 CFU per mL, approximately 4 CFU per mL would be generated after 3 h enrichment, which would not be sufficient to yield a positive result 100% of the time with the BAX® assay. An option for improving the theoretical assay sensitivity was to increase the relative volume of the enriched sample concentrate that was applied onto the detection kit for analyses. Usefulness of the immunomagnetic procedure to specifically capture organisms from a larger volume of the enriched sample was investigated, and compatibility of analyzing the IMS concentrate using the selected detection procedures was examined.

Initially, the performance of Dynabeads® anti-*E. coli* O157 (Dynal, Inc.) and *E. coli* O157 A-Beads™ (Aureon Biosciences) for recovering *E. coli* O157:H7 was evaluated. *E. coli O157:H7* (ATCC 35150) was used at a spike dose of $1 \times 10^4$ CFU with two sample volumes (1 mL and 10 mL), two diluents (PBS and TSB) and two different volumes of beads when processing 10 mL samples (100 μL vs. 200 μL). Following the manufacturer's instructions and using 20 μL of beads per 1 mL sample volumes, mean recoveries of 83% ± 27% (n=5) were obtained with Dynabeads® whereas Aureon A-Beads™ demonstrated mean recoveries of 55% ± 11% (n=4). Evaluating the capacity of the IMS step to handle larger sample volumes (i.e., 10 mL) was considered a useful adaptation to achieve the sensitivity required to yield a positive result with low levels of contamination. Extrapolation from the manufacturers directions suggested that 200 μL of beads would be required for a 10 mL sample volume, which would incur a cost of approximately $18 per test for the IMS component alone. To reduce the cost associated with performing the scaled up IMS, organism recovery efficiencies using 100 μL and 200 μL of Dynabeads® anti-*E. coli* O157 were compared in 10 mL volumes of PBS or TSB. The results are summarized in Table 3.3. Although using 100 μL of Dynabeads® in 10 mL samples (PBS or TSB) indicated mean recoveries that were marginally lower than those obtained with 200 μL of beads, assessing the variation associated with these recoveries indicated these differences were not significant.

**Table 3.3**
***E. coli* O157:H7 recovery efficiencies comparing sample size, bead volume and diluent**

| **PBS** | | **TSB** | |
|---|---|---|---|
| 100 μL beads (n=22) | 200 μL beads (n=3) | 100 μL beads (n=17) | 200 μL beads (n=3) |
| 81% ±17% | 113% ± 16% | 81± 12% | 101% ± 8% |

## Effect of Incubation Temperature on IMS

Both Dynabeads® and Aureon *E. coli* O157 A-Beads™ were evaluated using the manufacturer recommended incubation temperature (room temperature) and an elevated incubation temperature (42°C), which had previously been determined to be ideal for *E. coli* O157:H7 propagation in TSB. Analysis was performed using 10 mL sample volumes (either PBS or TSB), an *E. coli* O157:H7 (ATCC 35150) pike dose of $1x10^4$ CFU and an IMS beads volume of 100 μL per sample.

**Table 3.4**
***E. coli* O157:H7 IMS recovery efficiencies at room temperature versus 42°C.**

| **Matrix** | **Temperature** | **Dynabeads® anti-*E.coli* O157** | **Aureon *E.coli* O157 A-Beads ™** |
|---|---|---|---|
| PBS | Room Temperature | 81±17% (n=22) | 41±17.5% (n=11) |
| | 42°C | 71±14% (n=30) | 25±11.5% (n=5) |
| TSB | Room Temperature | 81±12% (n=17) | 37±11% (n=6) |
| | 42°C | 111±46.5% (n=20) | NA |

NA = Not Analyzed

The data presented in Table 3.4 using 10 mL sample volumes further validates the initial studies using 1 mL sample volumes in that significantly higher *E. coli* O157:H7 recoveries were obtained with Dynabeads® than Aureon beads, irrespective of the organism suspension medium or incubation temperature. Compared to room temperature, performing the IMS capture step at 42°C yielded a marginal reduction in mean organism recoveries (81% ± 17% compared to 71% ±

14%) in PBS using Dynabeads®. When TSB was used as the suspension medium, a marked increase in organism recovery efficiency was observed at 42°C (111% ± 46.5%) compared to room temperature (81% ± 12). Using TSB as the suspension medium *E. coli* O157:H7 recoveries were frequently over 100%. Although this could occur as a result of the initial spike doses being greater than had been anticipated, this was unlikely due to routine spike dose verification by spread plate enumerations. A more probable reason may have been that *E. coli* O157:H7 incubation in TSB at 42°C led to amplification of organisms and IMS beads did not impede growth of the target organisms. Should this be the case, it could potentially afford further streamlining of the procedure by simultaneously enriching and IMS capturing *E. coli* O157:H7. This phenomenon was investigated and has been discussed in a later section.

**Matrix Effects on IMS Recoveries**

The impact of different source water matrices on IMS recovery was examined during this study. IMS recovery efficiency in river water was 101% ± 26% compared to a recovery efficiency of 109% ± 27.5% in ground water. These recoveries are similar to those observed with organisms spiked into TSB (Table 3.4; 111% ± 46.5%), indicating that these environmental matrices had no significant effect on recovery of *E. coli* O157:H7 using IMS.

**Effect of Non-Target Organisms on IMS Recovery**

The capture of non-target organisms by the IMS step was evaluated and these cross-reactivity results have been presented in Table 3.5. TSA is a non-specific amplification media and it was expected that colony counts with TSA plates would yield the highest percentage increase in organism numbers if carryover of non-target organisms occurred following IMS. However, no significant increase in percentage recovery was observed in samples containing both non-target organisms and *E. coli* O157:H7. Using more selective media (sMAC plates and CHROMagar™ plates) demonstrated similar patterns. Where only non-target organisms were subjected to the IMS step (data not shown), majority (approximately 98%) of the non-target organisms were not captured, indicating low cross reactivity of the antibodies on the IMS beads for the panel of non-target organisms that was examined. The remaining 2% of non-target organisms that were carried into the final IMS concentrate may have occurred due to low levels of antibody cross-reactivity. It is more likely that, during the magnetic separation stage, these organisms were captured within the pellet of IMS beads and non-specifically carried over into the final concentrates. Previously, Bukhari *et al.*, (1998) have reported the occurrence of such a phenomenon for *Cryptosporidium parvum* oocysts subjected to IMS.

**Table 3.5**
**Recovery of spiked *E. coli* O157:H7 and non-target organisms following IMS**

| Spiked organisms | TSA Medium | sMAC Medium | CHROMagar™ Medium |
|---|---|---|---|
| *E. coli* O157:H7 | 101 ± 48% (n=3) | 93 ± 33% (n=3) | 176% (n=1) |
| *E. coli* O157:H7/non-O157 organisms | 115 ± 45% (n=3) | 112 ± 42% (n=3) | 140% (n=1) |

**Evaluating Simultaneous Enrichment and Capture of *E. coli* O157:H7**

Previously data suggested that the presence of IMS beads may not interfere with *E. coli* O157:H7 propagation dynamics at 42°C. This led to the possibility of simultaneous enrichment and IMS capture of *E. coli* O157:H7, which was considered useful for several reasons. Firstly it would help to avoid separate enrichment and IMS steps and potentially reduce the sample processing time by one hour. Secondly a modification of this nature would streamline the procedure, making it user-friendly and thirdly, it would avoid the unnecessary transfer of pathogenic organisms and thereby reducing the risk of accidental infections or potential losses from either sample transfer or spillages. Separate enrichment (3 h for BAX® assay and 6 h for Reveal® assay) followed by a separate one hour IMS step was compared with a simultaneous enrichment/capture procedure intended for the BAX® assay (total time 3 h) and Reveal® assay (total time 6 h). The results are presented in Table 3.6 and indicated that adding IMS beads to the enrichment broth approximately one hour prior to completion of the enrichment step did not significantly inhibit organism propagation. Furthermore, these data supported earlier data (Table 3.4) in that IMS performance was not deleteriously impacted when organism capture was conducted at a temperature (42°C) rather than room temperature. As a result the simultaneous enrichment and capture refinement was adopted in the subsequent protocol.

**Table 3.6**
**Comparison of two procedures for organism propagation and capture.**

| Initial Spike Dose (CFU/mL) | Separate Enrichment & Capture (4 h) (CFU/mL) | Simultaneous Enrichment & Capture (3 h) (CFU/mL) | Separate Enrichment & Capture (7 h) (CFU/mL) | Simultaneous Enrichment & Capture (6 h) (CFU/mL) |
|---|---|---|---|---|
| 1.0 | 174 ± 107.5 (n=20) | 112 ± 97 (n=32) | NA | NA |
| 0.5 | NA | NA | $2.8 \times 10^4 \pm 2.0 \times 10^4$ (n=10) | $7.4 \times 10^4 \pm 3.5 \times 10^4$ (n=10) |

NA = Not Analyzed

## PRELIMINARY TRIALS TO DETERMINE OVERALL METHOD SENSITIVITY

Based on the simultaneous enrichment and IMS capture format described in the previous section, preliminary sensitivity experiments were conducted using the overall assay and a range of initial spike doses. Data from these trials have been presented in Table 3.7.

**Table 3.7**
**Preliminary sensitivity measurements**

| Initial Spike Dose (CFU per mL) | CFU per mL (3 h) | % positive# by BAX® | CFU per mL (6 h) | % positive # by Reveal® |
|---|---|---|---|---|
| 1.0 | 112 ± 97 (n=32) | 100 | NA | NA |
| 0.5 | 32 ± 29 (n=24) | 100 | $7.5 \times 10^4 \pm 3.5 \times 10^4$ (n=10) | 100 |
| 0.25 | 13 ± 12 (n=15) | 47 | $4.3 \times 10^4 \pm 4.0 \times 10^4$ (n=14) | 100 |

NA = Not Analyzed
#=Estimates based on plate counts

Using a three hour enrichment and capture step and analysis of 10% of the concentrate indicated that an initial spike dose of either 1 CFU per mL or 0.5 CFU per mL would likely generate sufficient organism numbers (i.e., >10 CFU) to result in a 100% positive rate if BAX® was employed as the detection procedure, whereas a dose of 0.25 CFU per mL demonstrated only 47% of the samples contained sufficient organism numbers to yield a signal with the BAX® assay. The six hour enrichment/capture indicated that sufficient organisms would be recovered to generate a signal with the Reveal® assay using all the tested initial spike doses, including 0.25 CFU per mL. As only 10% of each sample concentrate was utilized for predicting the sensitivity of the overall procedures, the potential existed for improving overall method sensitivity by subjecting a larger proportion of the concentrate to analysis by either the BAX® assay or Reveal® assay. For example, increasing the equivalent volume in the IMS concentrate from 1 mL to 10 mL could potentially help to improve the method sensitivity 10 fold. Also analysis of a larger proportion of the concentrate could provide sufficient method sensitivity to detect *E. coli* O157:H7 numbers that would likely be encountered in contaminated water samples. Another option was to increase the original sample volume that would be processed by the procedure; however to achieve this, one or more samples concentration procedures had to be employed. The two concentration methods evaluated in this study were centrifugation and membrane filtration.

## EVALUATION OF CONCENTRATION METHODS

### Centrifugation

Initial centrifugation trials using 50 mL sample volumes indicated that there was a substantial difference between the recovery of *E. coli* O157:H7 in MQ water (4 ± 2 % recovery, n=2) and 0.2X TSB (119 ± 18% recovery, n=2) using a spike dose of $1 \times 10^4$ CFU. Decreasing the *E. coli* O157:H7 spike dose to 100 CFU reduced the average recovery to 78± 3% (n=2) in TSB.

Using a larger volume of MQ water (200 mL sample spiked with $1 \times 10^4$ CFU) and supplemented to a single strength (1 X) concentration in TSB yielded recoveries of 124 ± 21% (n=8). Impact of water sample matrices on recoveries with centrifugation was examined by using 200 mL volumes of autoclaved river water spiked with $1 \times 10^4$ CFU of *E. coli* O157:H7. Centrifugation with and without final concentrations of 1X and 2X TSB was performed to evaluate the effect of increasing media concentrations on organism recovery. Results are presented in Table 3.8.

**Table 3.8**
**Matrix effects on recovery efficiency with centrifugation**

| | **Percentage Recovery Efficiency (%)** | | |
|---|---|---|---|
| River Water | No TSB | 1X TSB | 2X TSB |
| OH | 66 | 69 | 77 |
| MO | 40 | 52 | 26 |
| WA | 22 | 42 | 33 |
| Average | 43± 18 | 54± 11 | 45± 23 |

For three untreated (raw) water sites, addition of 1X TSB improved recovery of *E. coli* O157:H7 whereas 2X TSB increased recovery in only one (OH) of these three sites. It is unknown why addition of TSB would improve organism recovery. Perhaps the adhesion of organisms to particulate matter in the TSB and/or the effect of TSB on the organism surface charges aided effective bacterial pelleting during centrifugation. Despite 1X TSB yielding a beneficial effect on organism recovery following centrifugation of spiked matrix samples, the average recoveries in spiked river water samples (54± 11%; Table 3.8) were substantially lower than recoveries in spiked 1X TSB-MQ water samples (124 ± 21%; discussed earlier). This may be due to some interaction of *E. coli* O157:H7 with particles in the source water matrices, such as adhesion or adsorption, leading to organism clumping or uneven distribution of the bacteria. To remedy this potential problem, and further improve recoveries from centrifuged source water pellets, employment of homogenization of the centrifuged source water pellets was tested but reduced average recovery of the three source waters from 50% to 29%. Based on these observations, homogenization of source water pellets was abandoned.

**Membrane Filtration (MF)**

Membrane filtration was examined as an alternative procedure for concentrating *E. coli* O157:H7 from 200 mL samples of MQ water and three spiked ($1 \times 10^4$ CFU *E. coli* O157:H7) raw water matrices. Recoveries ranged between 32% and 75% for the raw water matrices, whereas MQ water yielded a recovery of 67%. These MF recoveries were higher than those achieved with centrifugation (except when supplemented with TSB).

***Evaluating MF techniques***

To identify sources of organism loss during membrane filtration, MF was performed using three different approaches. Comparison was conducted between direct application of the spike dose to the filter (best scenario), direct application followed by filtration of 200 mL water samples (worst scenario) and filtration of the pre-spiked 200 mL water sample in its entirety (normal procedure). While the last method is only relevant for concentrating environmental water samples, the best and worst case scenarios were set up to help elucidate factors responsible for organism loss during centrifugation. In addition to the comparison of these three filtration scenarios, two different lots of filters and two different elution times were evaluated. The results are shown in Table 3.9

**Table 3.9**
**Mean *E. coli* O157:H7 recovery efficiencies using three types of filtered spiked samples**

| | Filter Lot # 1422S 10 min Elution | Filter Lot # 1213S 10 min Elution | Filter Lot # 1422S 2 h Elution |
|---|---|---|---|
| Direct application of spike dose | 64±38% (n=7) | 57±22% (n=2) | 51±0.5% (n=2) |
| Direct application followed by filtration of 200 mL MQ water | 41±30% (n=7) | 39±7% (n=2) | 43±5% (n=2) |
| Filtration of spiked 200 mL MQ water | 50±25% (n=7) | 31±12% (n=2) | 66±39% (n=2) |

A comparison of two filter lots and two elution times (Table 3.9) indicated no significant difference between filter lots or increasing the elution time from 10 min to 2 h. Comparisons between direct application of the spike dose followed by elution (64 ± 38%) and filtration of spiked 200 mL samples (50 ± 25%) indicated no significant differences. These data suggest that organism penetration through the filter was unlikely to be of concern; however, elution of the spike dose from filters was a likely limiting factor in organism recovery. Although increasing the elution time from 10 min to 2 h did improve mean organism recovery where filtration of spiked 200 mL samples was performed, the variability in recovery indicated differences were not significant. Although additional comparisons between various elution intervals may help optimize the elution recoveries further, current data indicated there was minimal fluctuation in organism recovery when different filter lots were evaluated. Future studies using polycarbonate track etched membrane (PCTE) filters, which have minimal depth, may be more suited to further improving the organism recovery. However, their lack of depth compared to the mixed cellulose ester filters, may make PCTE filters prone to blockages, especially when filtering untreated raw water samples.

### Concentration, Propagation and IMS

Membrane filtration and centrifugation as part of the overall enrichment and detection method was examined using several source water (four river and three ground) and finished water matrices (Table 3.10).

**Table 3.10**
***E. coli* O157:H7 recovery efficiencies by membrane filtration and centrifugation combined with enrichment and IMS**

| Matrix | Membrane Filtration Mean % Recovery ± SD | Centrifugation Mean % Recovery ± SD |
|---|---|---|
| MQ water | 43 ± 37% (n=22) | 73 ± 6% (n=4) |
| Raw water | 67 ± 39% (n=40) | 62 ± 24% (n=3) |
| Ground water | 116 ± 56% (n=6) | NA |
| Finished water | 44 ± 47% (n=12) | 111 ± 44% (n=7) |

NA = Not Analyzed

In spiked MQ water or finished water, centrifugation followed by enrichment and IMS resulted in greater recovery efficiencies than membrane filtration. In contrast, membrane filtration yielded slightly higher mean recoveries in raw water than centrifugation. Based on the variability in the recovery efficiencies, performance of both these concentration procedures were considered to be similar and membrane filtration was selected for incorporation into the overall method, primarily as this method is widely available to water utility labs, and these labs typically do not have access to the required centrifugation equipment. Also during centrifugation, the method required 10X TSB dilution to 1X TSB in the original 200 mL sample intended for centrifugation. While this suspension was useful for organism recovery, the 10X TSB diluted to 1X TSB was not as effective a medium for organism propagation as single strength (1X TSB). Poorer enrichment in 10X TSB when diluted to single strength TSB is unclear. It is possible that autoclaving the 10X TSB suspension burnt a portion of the media, thus reducing the metabolizable nutrient concentration below 10X. In turn, dilution of this media to single strength generated a suspension with less than 1X of metabolizable TSB. As environmental samples usually contain low levels of *E. coli* O157:H7, it was considered prudent to opt for a more effective enrichment procedure then a marginally better concentration procedure coupled with a 2 to 5 fold inferior enrichment.

## Effect of Non-Target Organisms on Overall Enrichment and IMS

Based on the previous optimization studies, the overall method selected for further evaluations consisted of membrane filtration, enrichment using TSB and simultaneous capture using IMS. This procedure was analyzed using various concentrations of non-target organisms in conjunction with low spike doses (0.5 –1.0 CFU per mL) of *E. coli* O157:H7. This was intended to determine whether interference from various concentrations of non-target organisms would affect the sensitivity of the overall method for isolation and detection of *E. coli* O157:H7. The effect of non-target organisms on method sensitivity was evaluated by comparing the number of presumed O157 colonies (based on morphology) in samples containing the mixture of non-target organisms to that of samples that contained *E. coli* O157:H7 only. Therefore, on sMAC plates only white colonies were counted, and on CHROMagar™ O157 plates only mauve colonies were counted. Little or no morphological differences were observed between *E. coli* O157:H7 and the non-target organisms used in this study when samples were plated on TSA plate. Therefore, analysis on TSA plates was excluded. The results are shown in Table 3.11.

**Table 3.11**
**Impact of non-O157 organisms on overall method sensitivity**

| ***E. coli* O157 :H7 spike dose** | **0.5 CFU/mL** | | **1 CFU/mL** | |
|---|---|---|---|---|
| | sMAC Medium | CHROMagar™ Medium | sMAC Medium | CHROMagar™ Medium |
| 100 CFU mixture of non-O157 organisms | +84± 67% (n=2) | +111± 19% (n=3) | +29% (n=1) | -40% (n=1) |
| 1,000 CFU mixture of non-O157 organisms | +36% (n=1) | +400% (n=1) | -50± 6% (n=2) | -51% (n=1) |
| 10,000 CFU mixture of non-O157 organisms | -87% (n=1) | +125± 156% (n=2) | -57± 23% (n=4) | -68% (n=1) |

+ values indicate percentage amplification
- values indicate percentage reduction

With an initial spike dose of 0.5 CFU per mL of *E. coli* O157:H7 and mixtures consisting of up to 1,000 CFU of non-target organisms, analyses by both sMAC and CHROMagar plates indicated no negative impact on the growth rate of *E. coli* O157:H7 by the presence of non-target organisms. However, when the non-target spike doses were increased to 10,000 CFU, sMAC plates indicated 87% reduction in *E. coli* O157:H7 growth rate, whereas enumeration on CHROMagar™ plates indicated no interference with *E. coli* O157:H7. Increasing the *E. coli* O157:H7 spike dose to 1 CFU per mL and adding various concentrations of non-target organisms demonstrated reducing growth rates of *E. coli* O157:H7 when using CHROMagar™ plates. sMAC demonstrated similar patterns at the two higher concentrations of non-target organism. Absence of interference using lower spike doses of *E. coli* O157:H7 was consistent with previous data indicating low cross reactivity of individual components of the method. In contrast, the growth rate inhibition observed when using increased spike doses of *E. coli* O157:H7 was probably an aberration. These data confirmed that the panel of non-target organisms examined do not pose significant interference to the methods selected for isolation of *E. coli* O157:H7 from source and finished water samples.

## CONFIRMATION PROCEDURES

### Latex Agglutination

The Oxoid test, which identifies only members of the *E. coli* O157 serogroup, produced a positive agglutination reaction with $1x10^7$ CFU of *E. coli* O157:H7 (ATCC 35150). Concentrations of $\leq 1x10^6$ CFU of *E. coli* O157:H7 did not result in positive agglutination reactions. Positive and negative controls gave appropriate reactions in this test. The Murex

Wellcolex test uses two separate antibodies to test for the *E. coli* O157 and H7 antigens. When the O157 portion of the test was evaluated a positive agglutination reaction was observed with $1x10^7$ CFU of *E. coli* O157:H7 (ATCC 35150). The H7 portion of the test required $1x10^8$ CFU to obtain a positive agglutination reaction. Lower concentrations of *E. coli* O157:H7 (i.e., $1x10^6$ CFU for the O157 test and $1x10^7$ CFU for the H7 test) did not result in positive agglutination reactions. Positive and negative controls gave appropriate reactions in this test.

## SINGLE LABORATORY VALIDATION OF OVERALL METHOD

### Overall Method Sensitivity

The overall methods (membrane filtration, enrichment, IMS and detection by Reveal® or BAX®) developed during this project were initially subjected to sensitivity measurements in a single laboratory validation, and these data have been presented in Table 3.12.

**Table 3.12**
**Single laboratory validation conducted at the Belleville laboratory**

| Spike Dose CFU/mL | Matrix | CFU 3 h; 42°C* | Percentage BAX® Positive | CFU 6 h; 42°C* | Percentage Reveal® Positive |
|---|---|---|---|---|---|
| 5.0 | MQ Water | 13000 | 100 (1/1) | NA | 100 (1/1) |
| | Tap Water | 16000 | 100 (1/1) | NA | 100 (1/1) |
| | Raw Water | NA | 100 (1/1) | NA | 100 (1/1) |
| 0.5 | MQ Water | 1960 | 100 (1/1) | NA | 100 (1/1) |
| | Tap Water | 2106±2004 n=5 | 100 (5/5) | $1.1x10^6 \pm 0.9x10^5$ n=2 | 100 (2/2) |
| | Raw Water | NA | 100 (1/1) | NA | 100 (1/1) |
| 0.25 | MQ Water | 2240 | 100 (1/1) | NA | 100 (1/1) |
| | Tap Water | 1200 | 100 (1/1) | NA | 100 (1/1) |
| | Raw Water | NA | 100 (1/1) | NA | 100 (1/1) |
| 0.05 | MQ Water | 445±393 n=26 | 100 (4/4) | $9.2x10^4 \pm 2.1x10^4$ n=8 | 100 (7/7) |
| | Tap Water | 320±281 n=19 | 60 (6/10) | $3.2x10^5 \pm 5.7x10^4$ n=4 | 100 (4/4) |
| | Raw Water | 396±275 n=46 | 85 (11/13) | NA | 100 (5/5) |

* Spread plate enumerations

NA = Not Analyzed

Using all three matrices (MQ water, tap water and raw water), the Reveal® assay yielded 100% positive reactions at spike doses ranging from 5.0 CFU per mL to 0.05 CFU per mL. With the BAX assay, reactions in all three water types were positive at spike doses of 5.0 CFU per mL, 0.5 CFU per mL and 0.25 CFU per mL. However, at the lowest spike dose (0.05 CFU per mL), 100% positive BAX® reactions were noted only in spiked MQ water samples, whereas 60% and 80% of reactions were positive from tap and raw water matrices, respectively. The increased frequency of negatives with the BAX® assay occurred despite the spread plate enumerations indicating the presence of a sufficient numbers of organisms (>320 CFU) to generate a positive BAX® reaction. One possible reason for this could be carryover of inhibitory compounds present in the tap (i.e., coagulants and/or flocculants) and raw (i.e., humic and/or fulvic acids) waters. A second cause may be carryover of immunomagnetic beads into the BAX® assay, causing interference in the amplification process during the PCR reactions. Based on these observations, it became necessary to evaluate how the BAX® assay sensitivity may be impacted by contamination of the BAX® PCR reactions with IMS beads.

**Impact of IMS Bead Carryover on BAX® Assay Sensitivity**

Following IMS, the effect of carryover of trace amounts of IMS bead on the detection sensitivity of the BAX® assay was assessed (Table 3.13).

**Table 3.13**
**Effect of IMS bead carryover on BAX® assay sensitivity**

| Spike Dose per BAX® reaction | Volume Dynal IMS Beads per BAX® reaction (µL) | BAX® Result # positive/ # replicates (%) |
|---|---|---|
| 10 | 0 | 4/7 (57%) |
| | 20 | 0/5 (0%) |
| | 50 | 1/7 (14%) |
| 25 | 0 | 2/2 (100%) |
| | 50 | 0/2 (0%) |
| 50 | 0 | 2/2 (100%) |
| | 50 | 1/2 (50%) |
| 100 | 0 | 2/2 (100%) |
| | 50 | 2/2 (100%) |

At the lowest spike dose (10 CFU) and in the absence of beads, while some negative reactions occurred, the frequency of negatives increased following addition of IMS beads. For example, using a spike dose of 10 CFU with 20 and 50 µL of beads generated 0% and 14% BAX® positive reactions respectively, compared to 57% positive reactions in the absence of beads. The results (Table 3.13) and (Figure 3.1) indicated that contamination of the BAX® reaction with up to 50 µL of IMS beads could be responsible for four to ten fold decrease in BAX® assay sensitivity for detection of low levels of *E. coli* O157:H7. Despite the internal positive control indicating successful PCR reactions for the samples containing IMS beads, inhibition of PCR reactions for the target organism cannot be excluded. There is also a possibility that presence of IMS beads prevented effective lysis of the *E. coli* O157:H7, thereby reducing the amount of target sequence available for the BAX® reaction. Another explanation may be that samples with IMS beads contained less than the target 10 CFU per reaction. Although any one of these factors may have influenced the outcome of the BAX® reactions, a convenient resolution to circumvent this problem was to exercise care during transfer of the lyzed supernatant onto the BAX® tablet in order to minimize carryover of IMS beads.

### Modification of the Method Incorporating Reveal® Detection

During the final stages of the single lab validation the protocol developed for rapid isolation and detection of *E. coli* O157:H7 from various water types was finalized. The most notable change made to the procedure was the need to adjust the incubation temperature from 42°C to 35°C for those labs using the Reveal® detection device. This was necessary as the participating water utility labs did not have ready access to incubators operating at 42°C due to the need to maintain their available incubators at 35°C for compliance testing.. This modification in the Reveal® procedure was evaluated using various spike doses of *E. coli* O157:H7 suspended in finished water samples (Table 3.14).

**Table 3.14**
**Evaluation of modified Reveal® detection method**

| Incubation Temperature | Spike Dose CFU per mL | Reveal® Result # positive/ # replicates | Average CFU/ Reveal® Device |
|---|---|---|---|
| 35°C | 0.05 | 4/4 | $4.1x10^4$ |
| | 0.50 | 2/2 | $1.25x10^5$ |
| | 1.0 | 1/1 | $4.0x10^5$ |
| 42°C | 0.05 | 4/4 | $3.23x10^5$ |
| | 0.50 | 2/2 | $1.0x10^6$ |
| | 1.0 | 1/1 | $1.0x10^7$ |

Results indicated propagation of *E. coli* O157:H7 was less effective at 35°C than at 42°C (1 to 2 logs lower); however, incubation at both temperatures generated sufficient numbers of organisms to produce a positive result by the Reveal® device. These data indicated that use of 35°C rather than 42°C would not compromise the overall method sensitivity when using an initial spike dose of >0.05 CFU/mL. While it would be preferable for organism enrichment to be performed at 42°C, amplification of organisms at 35°C would not compromise overall method sensitivity and was a viable option for use in the round-robin study.

## MULTI-LABORATORY ROUND ROBIN

### *E. coli* O157:H7 Stability During Transit

In the start up phase of the round robin, it was noted that a significant reduction in *E. coli* O157:H7 organism numbers occurred during the shipment of the spike doses. While other researchers have reported a 1-2 log reduction after 91 days at 8°C (Wang and Doyle, 1998), in this study similar reductions occurred after an overnight shipment on ice. To compensate for this, all subsequent shipments to participating labs contained a 1 log higher number of organisms than the desired target number intended for evaluation in the round robin. In addition, all round robin labs performed additional spread plate enumerations on the shipped spike doses that confirmed a reduction in *E. coli* O157:H7 numbers of 1 to 1.5 logs.

## Overall Method Using the Reveal® Detection Assay

A total of nine American Water utility labs participated in round robin testing of an overall method for rapid detection of *E. coli* O157:H7 using the Reveal® detection device. These labs analyzed both finished and raw water samples using eight 'blind' spike doses per water matrix. The spike doses ranged from 1 CFU to 200 CFU and included one negative (unspiked) control dose. Results from three of the nine participating labs indicated laboratory contamination of the negative (unspiked) control using finished water. Data from these labs were considered invalid. Results from the remaining six labs, using finished water, are presented in Table 3.15 and indicated 100% positive rate with the overall method when 20 CFU or higher were spiked per 200 mL sample volumes. Even with the initial spike dose of 1 CFU per 200 mL (or 0.005 CFU per mL), a 33% positive rate was determined for finished water using the Reveal® assay.

**Table 3.15**
**Multi-laboratory validation using Reveal® assay in finished water**

| | % Positive Reveal® Reactions | | | | |
|---|---|---|---|---|---|
| **Spike Dose** | **1 CFU** | **10 CFU** | **20 CFU** | **100 CFU** | **200 CFU** |
| Lab A | 33 | 100 | 100 | NA | NA |
| Lab B | 33 | 100 | 100 | NA | NA |
| Lab C | 33 | 100 | 100 | NA | NA |
| Lab D | NA | 0 | NA | 100 | 100 |
| Lab E | NA | 100 | NA | 100 | 100 |
| Lab F | NA | 100 | NA | 100 | 100 |
| Mean | 33 | 80 | 100 | 100 | 100 |

NA = Not Analyzed

Using the raw water matrix five of six participating labs obtained a positive Reveal® reaction in the unspiked raw water. All six labs detected a positive signal in all the spike doses evaluated. While obtaining positive reactions in the unspiked raw water may have been due to cross-reactivity of the detection procedure, it is also possible this was due to presence of indigenous *E. coli* O157:H7 in the raw water.

## Overall Method Using the BAX® Detection Assay

A total of six laboratories participated in the round robin evaluation of the BAX® system for detection of *E. coli* O157:H7. One lab was unable to continue due to staffing issues and another failed to submit data until after the project was completed. Their data was considered void. The remaining four labs analyzed both finished and raw water samples using eight 'blind' spike doses per water matrix. The spike doses ranged from 1 CFU to 200 CFU and included one negative (unspiked) control dose. All four participating labs obtained negative BAX® results for the negative (unspiked) controls in both finished and raw water. Results from the spiked samples are shown in Tables 3.16 and 3.17.

**Table 3.16**
**Multi-laboratory validation using BAX® assay in finished water**

| | % Positive BAX® Reactions | | | | |
|---|---|---|---|---|---|
| **Spike Dose** | **1 CFU** | **10 CFU** | **20 CFU** | **100 CFU** | **200 CFU** |
| Lab J | 0 | 100 | 100 | NA | NA |
| Lab K | NA | 100 | NA | 100 | 100 |
| Lab L | NA | 0 | NA | 0 | 100 |
| Lab M | NA | 67 | NA | 100 | 100 |
| Mean | 0 | 64 | 100 | 67 | 100 |

NA = Not Analyzed

Using the finished water matrix, the average percentage BAX® positives ranged from 0% (one lab that evaluated 1 CFU) to 100% (20 CFU and 200 CFU). At a spike dose of 100 CFU, the average positive rate for the four labs was 67%. This average positive rate may be an aberration as three of four labs obtained 100% positive reactions in spiked samples whereas one lab detected no positive reactions (Table 3.16). In contrast to the finished water samples, spike doses of 20 CFU or greater (per 200 mL raw water samples) consistently yield 100% BAX® positives during analysis by the round robin labs. Even with the initial spike dose of 1 CFU per 200 mL (or 0.005 CFU per mL), a 33% positive rate was determined for raw water using the BAX® assay (Table 3.17).

**Table 3.17**
**Multi-laboratory validation using BAX® assay in raw water**

| | % Positive BAX® Reactions | | | | |
|---|---|---|---|---|---|
| **Spike Dose** | **1 CFU** | **10 CFU** | **20 CFU** | **100 CFU** | **200 CFU** |
| Lab J | 33 | 100 | 100 | NA | NA |
| Lab K | NA | 67 | NA | 100 | 100 |
| Lab L | NA | 100 | NA | 100 | 100 |
| Lab M | NA | 67 | NA | 100 | 100 |
| Mean | 33 | 82 | 100 | 100 | 100 |

NA = Not Analyzed

Multi-laboratory round robin evaluations indicated the Reveal® assay was more reproducible and sensitive than the BAX® detection system when examining finished water samples. These data were in agreement with the single laboratory validation data. For example, at a spike dose of 10 CFU per 200 mL samples, five of the six external labs determined the presence of a positive Reveal® signal in 100% of the samples. In contrast to this, the BAX® assay indicated 60% positive in the single laboratory validation compared to 64% in the external round robin trials. When using raw water samples, five of the six external labs reported a positive signal in unspiked raw water samples using the Reveal® method. While this may indicate laboratory contamination of the unspiked raw water sample, it may also be due to indigenous *E. coli* O157:H7 in the raw water or detection of a signal from cross-reactivity of the Reveal® assay. None of the laboratories analyzing raw water samples with the BAX® assay detected a signal in unspiked samples. The latter may be due to higher specificity of the BAX® assay reducing the likelihood of cross-reactivity in raw water samples containing a larger variety of non-target organisms. Further trials are necessary to elucidate whether cross-reactivity issues exist with the Reveal® assay that were not readily identified with the limited species of non target organisms used in the initial specificity testing studies conducted during this project. Until this is resolved, perhaps it would be prudent to use the BAX® assay only for analysis of raw water samples, whereas both the BAX® and Reveal assay could serve as preliminary screening tools for analysis of finished water samples.

# CHAPTER 4
# RAPID DETECTION OF *E. COLI* O157:H7 IN WATER SAMPLES

Based upon the investigations reported in the previous chapters, rapid detection of *E. coli* O157:H7 in environmental water samples can be accomplished by concentration of 200 mL (raw, finished or ground water) samples by membrane filtration (MF). Following elution of the sample concentrates, enrichment of recovered organisms is performed in Tryptic Soy Broth (TSB), purification by immunomagnetic separation (IMS) and detection using the Reveal® and/or BAX® detection devices. It is possible to perform organism enrichment for six hours at either 35°C or 42°C where detection is performed with the Reveal® assay. The three hour enrichment prior to detection using the BAX® assay can only be performed at 42°C.

Good laboratory practices and adherence to the described methods is necessary to ensure success of these detection methods.

## MATERIALS REQUIRED

### General Equipment and Reagent Requirements

- Vacuum source and vacuum manifold (Pall Gelman, Ann Arbor, MI)
- MicroFunnel™ membrane filter (Pall Gelman, Ann Arbor, MI) (0.45 µm, 47 mm)
- Incubator set at 35°C and/or 42°C (Precision Scientific, Chicago, IL)
- Pipet-Aid (Drummond Scientific Co., Broomall, PA)
- Disposable graduated serological pipets (VWR, West Chester, PA)
- Sterile disposable petri dishes, 100 mm diameter (VWR, West Chester, PA)
- L-shaped spreaders (Fisher Scientific, Pittsburgh, PA)
- Leighton tubes (Bellco Glass, Inc., Vineland, NJ)
- 1.5 mL Eppendorf microcentrifuge tubes (VWR, West Chester, PA)
- 15 mL and 50 mL conical centrifuge tubes (VWR, West Chester, PA)
- Forceps/Tweezers (VWR, West Chester, PA)
- Bunsen burner for flame sterilization
- Vortex mixer (Scientific Industries, Bohemia, NY)
- Autoclave for proper decontamination and disposal of used supplies
- Racks for 1.5 mL Eppendorf tubes, Leighton tubes and 50 mL conical tubes
- Bottles: 1 L for sample collection and 250 mL (sterile glass or plastic) for samples
- Pipetman (Rainin Instrument Co., Woburn, MA)
- Latex (or suitable material) gloves
- Protective clothing (lab coat) and eye protection (eye glasses, safety glasses or goggles)
- Dynal Biotech Sample Mixer (Dynal, Inc., Lake Success, NY)
- Dynal MPC®-1 magnet (Dynal, Inc., Lake Success, NY)
- Dynal MPC®-S magnet (Dynal, Inc., Lake Success, NY)
- *E. coli* O157:H7 (ATCC# 35150) (American Tissue Culture Collection, Manassas, VA)- Prepare stock culture of bacteria according to procedure described on vial.
- Dynabeads® anti-*E. coli* O157 (Dynal, Inc., Lake Success, NY)

- Milli-Q (MQ) Ultrapure Water (Millipore, Bedford, MA) or equivalent reagent grade water.
- Ethanol for sterilization (Sigma, St. Louis, MO)
- 10% Bleach or 70% Ethanol suitable for cleaning equipment and surfaces.
- 10% (w/v) Sodium thiosulfate - Add 10 g sodium thiosulfate ($Na_2S_2O_3$) per 90 mL MQ water. Mix until dissolved. Bring to final volume of 100 mL and autoclave for 15 min at 121°C and 15 psi, store at 4°C.
- Difco Tryptic Soy Broth (TSB) (Becton, Dickinson and Co., Sparks, MD) Follow procedure as specified on bottle using MQ water.
- Difco Tryptic Soy Agar (TSA) (Becton, Dickinson and Co., Sparks, MD) Follow procedure as specified on bottle using MQ water.
- Phosphate Buffered Saline (PBS) (Sigma Co., St. Louis, MO) Follow procedure as specified on bottle using MQ water.
- Tween-20 (Sigma Co., St. Louis, MO)
- IMS Wash Buffer - Add 50 µL of Tween-20 to 100 mL of PBS, mix thoroughly, store at 4°C.
- Sorbitol-MacConkey (sMAC) agar (Becton, Dickinson and Co., Sparks, MD)

### Equipment and Supplies Required for Reveal® Detection Method

- Hot plate for boiling water
- Nalgene floating microcentrifuge tube rack (VWR, West Chester, PA)
- Reveal® for *E. coli* O157:H7 Test (Neogen Corporation, Lansing, MI)

### Equipment and Supplies Required for BAX® Detection Method

- Heat blocks or water baths to be used at 37°C and 95°C
- Cold block chilled to 4°C
- BAX® *E. coli* O157:H7 assay (DuPont Qualicon, Wilmington, DE)
- BAX® System (DuPont Qualicon, Wilmington, DE)

## QUALITY CONTROL PROCEDURES

### Enumeration of *E. coli* O157:H7 Stock Cultures

1. Prepare a stock culture of *E. coli* O157:H7 (ATCC# 35150) in TSB (as described by ATCC).
2. Prepare serial dilutions in PBS (i.e., 1 mL of culture into a tube containing 9 mL of PBS, mix thoroughly and remove 1 mL; add this 1 mL to a second tube containing 9 mL of PBS, mix thoroughly and remove 1 mL; add this 1 mL to a third tube containing 9 mL of PBS, mix thoroughly and remove 1 mL, etc.). It is recommended that serial dilutions of the stock culture are prepared up to $10^{-8}$.
3. Spread plate 100 µL (in duplicate) of the last three dilutions (i.e., $10^{-6}$, $10^{-7}$, $10^{-8}$ ) onto TSA plates and incubate plates overnight at 37°C.

4. Enumerate colonies on the plates and determine the average count for all plated dilutions that contain 30 – 300 CFU.
5. Using the highest average count, adjust the numbers to account for the dilution factor and multiply the adjusted number by 10 to determine the concentration of the stock culture in colony forming units (CFU) per mL.

   **Example:** using a serial dilution of $10^{-7}$, duplicate 100 μL volumes plated on TSA plates indicated 120 CFU and 98 CFU. First determine the average count per 100 μL = (120 CFU + 98 CFU)/2 = 109 CFU. Secondly adjust the average count to account for the dilution factor (in this case $1x10^7$)=109 $x10^7$=$1.09x10^9$. As this concentration is based on plating 100 μL, adjust the count to CFU per mL by multiplying $1.09x10^9$ by 10 = $1.09x10^{10}$.

   Based on this example, the average concentration in the stock would be $1.09x10^{10}$ CFU per mL.

### Preparation of Inocula and Matrix Spikes

1. Using a pre-enumerated stock culture, calculate the appropriate volume of either stock culture or stock culture dilution that contains a spike dose ranging between 10-100 CFU.
2. Dispense 200 mL sub-samples of each water matrix intended for analysis into individual 250 mL sterile glass screw cap bottled. Loosely screw cap onto each bottle.
3. Using a pipette, add the predetermined spike volume to deliver a known number of *E. coli* O157:H7 into each 200 mL sub-sample of each water matrix.
4. Simultaneously dispense and spread the same calculated volume onto duplicate TSA plates, incubate overnight at 37°C and enumerate the colonies the following day.
5. Process the spiked matrix samples through the entire procedure (continue step 7 onwards) for both Reveal® and BAX® detection methods.

### Preparation of Method Blank (Negative Control)

Process 200 mL of unspiked MQ water through the entire procedure (continue step 7 onwards) for both Reveal® and BAX® detection methods.

### Preparation of Reveal® Device Positive Control (Required for Step 11 Reveal® Detection)

Using PBS, dilute an aliquot of a pre-determined stock culture of *E. coli* O157:H7 to prepare a dilution that contains $5.0x10^5$ CFU per mL.
Place the *E. coli* O157:H7 into a 1.5 mL Eppendorf tube.
Boil the positive control for 10 min and cool to room temperature.
Positive control aliquots may be prepared in advance and stored at 4°C for up to six months
When analyzing environmental samples using the Reveal® device, run one positive control per batch of samples by adding 100 μL of positive control to the sample port of an individual Reveal® device.

**Preparation of BAX® Lysis Reagent (Required for Step 10 BAX® Detection)**

Follow the instructions provided in the BAX® *E. coli* O157:H7 assay kit.
Prepare lysis reagent by diluting protease 1/80 in lysis buffer (i.e., mix 12.5 μL protease and 987.5 μL lysis buffer).
Lysis reagent may be stored at 2-8°C and used for up to two weeks from preparation date.

**Preparation of BAX® Device Positive Control (Required for Step 11 BAX® Detection)**

Using PBS, dilute an aliquot of a pre-determined stock culture of *E. coli* O157:H7 to prepare a dilution that contains $2.0 \times 10^4$ CFU per mL. Place the *E. coli* O157:H7 into a 1.5 mL Eppendorf tube. Verify *E. coli* O157:H7 concentration by spread plate enumeration on TSA plates.
Add 5 μL containing 100 CFU of *E. coli* O157:H7 to 200 μL of lysis reagent in a 1.5 mL Eppendorf tube.
Using pre-heated heat blocks, incubate the tube at 37°C for 20 min followed by 10 min at 95°C.
Cool the positive control sample to 4°C.
Add 50 μL of the heat treated positive control sample to hydrate one BAX® *E. coli* O157:H7 assay PCR tablet immediately before using the BAX® system. The positive control should not be prepared in advance, but must be processed simultaneously with test samples.

**Preparation of BAX® Device Negative Control (Required for Step 11 BAX® Detection)**

Add 50 μL of prepared lysis reagent only to a 1.5 mL Eppendorf tube.
Using pre-heated heat blocks, incubate the tube at 37°C for 20 min followed by 10 min at 95°C.
Cool the negative control sample to 4°C.
Use a 50 μL aliquot to hydrate one BAX® *E. coli* O157:H7 assay PCR tablet immediately before using the BAX® system. The negative control should not be prepared in advance, but must be processed simultaneously with test samples.

## REVEAL® DETECTION PROCEDURE

### Sample Collection and Storage

For each water matrix being analyzed for the presence of *E. coli* O157:H7, collection of a 1 L grab sample is recommended. A 200 mL sub-sample is analyzed for each water matrix. It is recommended that at least 20% of the total number of samples analyzed for each matrix are subjected to matrix spike analyses. The sample volume used for each matrix spike should be equivalent to the volume analyzed for the test samples (i.e., 200 mL sub-sample). For finished water (tap water) samples, residual chlorine must be neutralized by adding sodium thiosulfate to a final concentration of 0.01%.

Between sample collection and analysis, it is recommended that all samples are stored in a refrigerator (4°C).

**Sample Concentration**

1. Prepare the vacuum, vacuum manifold and Microfunnel™ disposable filter units. Ensure that the funnel portion of the unit is securely snapped into place as it can dislodge during shipment
2. Flame sterilize forceps or tweezers (used for transferring membrane filters from manifold to the 50 mL conical tubes). The forceps or tweezers require sterilization between each sample membrane.
3. Using an incubator set at 42°C (or 35°C), warm a sufficient volume of sterile TSB for the total sample load. A minimum of 10 mL of TSB is required for each sample.
4. Retrieve the water samples from the refrigerator and mix thoroughly by inversion.
5. Dispense 200 mL of each test water sample into individual sterile bottles and label with unique identification codes and designate samples as unspiked test samples.
6. Prepare matrix spike and method blank according to procedures described in QC section.
7. Filter each sample (unspiked, matrix spike or method blank) through individual Microfunnel™ filters, at a flow rate of approximately 100 mL per min.
8. Remove each membrane filter from its filter housing by snapping off the funnel portion of the unit. Use forceps to remove the membranes and place in individual pre-labeled 50 mL conical centrifuge tubes. Sterilize the forceps between membranes. The filter funnel and base can be discarded.
9. Add 10 mL of 1X TSB to each conical tube, cap tube and vortex samples at maximum speed for 1 min. Allow to stand for at least 10 sec and repeat vortex (1 min).
10. For each sample, transfer the filter eluant to an individual pre-labeled Leighton tube. Discard the conical tube and filter membrane.
11. Continue with the sample enrichment and IMS procedure.

**Sample Enrichment And IMS**

12. Place a Dynal Sample Mixer in an incubator set at 42°C (or 35°C).
13. Attach Leighton tubes to the Sample Mixer. Verify that the caps on the Leighton tubes are on securely.
14. Start the Dynal Sample Mixer, adjust rotational speed to 24 rpm and incubate samples for 5 h.
15. Remove the Leighton tubes from the Sample Mixer and add 100 μL of Dynabeads® anti-*E. coli* O157:H7 to each Leighton tube. (Paramagnetic beads will settle during shipment. Thoroughly resuspend the beads before use.)
16. Return Leighton tubes to the Sample Mixer and continue incubation at 42°C (or 35°C) for one additional hour (rotational speed 24 rpm).
17. During the last hour of enrichment, pre-label Eppendorf tubes, set out magnets, and prepare IMS wash buffer.
18. Remove the Leighton tubes from the Sample Mixer one at a time and place in the MPC®-1 magnets.
19. Ensure flat side of the Leighton tube is flush with the magnetic portion of the MPC®-1.
20. Gently rock the MPC-1 and Leighton tube from end to end (180 degrees motion) for 3 min.

21. Decant the supernatant, holding the Leighton tube in the magnet, with the magnet side being uppermost (i.e., pour the supernatant away from the magnet). Paramagnetic beads should form a visible pellet on the flat side wall of the Leighton tube, adjacent to the magnet. When removing the cap from the Leighton tube, ensure that the tube does not move/rotate in the MPC®-1.
22. Exercise appropriate infectious material handling precautions in discarding the supernatant. Autoclave any liquid waste before disposal.

*1st Wash*

23. Remove the Leighton tube from the MPC®-1 and resuspend the bead-bacteria complex using 0.8 mL of IMS wash buffer using a 1 mL pipet.
24. Gently perform repeated pipeting to resuspend the beads. Transfer the resuspended beads to a 1.5 mL Eppendorf tube pre-labeled with the sample ID.
25. Add an additional 0.5 mL of IMS wash buffer to the Leighton tube and gently rinse the inside of the Leighton tube. Add this wash to its respective Eppendorf tube.
26. Repeat steps 7-14 with all samples using the MPC®-1 magnet before continuing to step 16.
27. Place the Eppendorf tubes (containing 1.3 mL - 1.5 mL of the sample bead-bacteria complex) into MPC®-S magnets and slide the magnet strip into place. Align tubes such that the hinged side of the tube is flush with the magnet.
28. Gently rock the Eppendorf tubes in the MPC®-S from end to end (90 degree motion) for 3 min.
29. Keeping the Eppendorf tubes secured in the MPC®-S, gently open the tube caps and use a P1000 pipetman (or Pasteur pipet) to remove the supernatant. Avoid disturbing the bead-bacteria complex that has formed a pellet on the hinged side wall of the Eppendorf tube. Change pipet tips between samples to avoid cross-contamination. Discard the supernatant, exercising appropriate care for handling infectious material.

*2nd Wash*

30. Remove the magnetic strip from the MPC®-S (the tubes can remain in the holder). Add 1 mL of IMS wash buffer to the bead pellet in each tube, using a P1000 or 1 mL disposable pipet. Use a clean pipet tip (or disposable pipet) for each Eppendorf tube.
31. Cap all the Eppendorf tubes and gently rock from end to end (90 degrees motion) 5-10 times or until the bead pellet has been resuspended.
32. Insert the magnet into the MPC®-S and rock the entire assembly from end to end (90 degrees motion) for 3 min.
33. Keeping the Eppendorf tubes in the MPC®-S, gently open the tube caps and use a P1000 pipetman (or Pasteur pipet) to remove the supernatant. Avoid disturbing the bead-bacteria complex. Change pipet tips between samples to avoid cross-contamination. Discard the supernatant, exercising appropriate care for handling infectious material

### *3rd Wash*

34. Remove the magnetic strip from the MPC®-S (the tubes can remain in the holder). Add 1 mL of IMS wash buffer to the bead pellet in each tube, using a P1000 or 1 mL disposable pipet. Use a clean pipet tip or disposable pipet for each Eppendorf tube.
35. Cap all the Eppendorf tubes and gently rock from end to end (90 degrees motion) 5-10 times or until the bead pellet has been resuspended.
36. Insert the magnet into the MPC®-S and rock the entire assembly from end to end (90 degrees motion) for 3 min.
37. Keeping the Eppendorf tubes in the MPC®-S, gently open the tube caps and use a P1000 pipetman (or Pasteur pipet) to remove the supernatant. Avoid disturbing the bead-bacteria complex. Change pipet tips between samples to avoid cross-contamination. Discard the supernatant, exercising appropriate care for handling infectious material
38. Remove the Eppendorf tubes from the MPC®-S assembly and transfer to a regular microtube rack.
39. Add 100 μL of IMS wash buffer to each tube and resuspend the bead-bacteria pellet by vortexing for 10-30 sec.
40. Process all samples to step 39 before continuing with the Reveal® detection protocol.

### Reveal® Detection

1. Transfer 5 μL of the bead-bacteria complex from each sample to sterile 1.5 mL Eppendorf tubes and store this sub-sample at 4°C. This sub-sample can be used later to for confirmation of Reveal® positive samples, plated onto sMAC plates or for BAX® analysis.
2. Place the Eppendorf tubes containing the 95 μL of bead-bacterial concentrates from the samples in a floating microtube rack.
3. Place the rack in a beaker of boiling water and boil samples for 10 min.
4. Allow boiled test samples to cool for 5-10 min at 4°C or on ice.
5. Condensation may accumulate in the Eppendorf tube lid. Either shake the tubes in a downward motion or perform a brief centrifugation step (i.e. 10,000-12,000 x g; 10 sec) to collect residual liquid in the bottom of the tube.
6. Remove the appropriate number of Reveal® devices from the 4°C refrigerator and warm to room temperature (5 - 10 min). (Individual Reveal® devices will be required for all unspiked test samples, matrix spike, method blank, and Reveal® positive control).
7. Place the Eppendorf tubes containing the boiled samples into the MPC®-S. Ensure hinged side of each tube rests against the magnet.
8. Place the MPC®-S on a benchtop for 1- 2 min, without disturbing the samples.
9. For each sample, use a P200 pipetman (or Pasteur pipet) to carefully aspirate the 100 μL of supernatant. Take care to avoid drawing up the pelleted beads.
10. For each sample, dispense the entire supernatant into the sample port of an individual Reveal® test device.
11. Load the Reveal® positive control into the sample port of an individual Reveal® test device.
12. Allow all test and control samples to rest horizontally on a flat surface (i.e., bench top) and wick up the display portion of the Reveal® device.

13. Observe and record test results after 15 min. (Pink colored bands will form within minutes). Bands appearing after 20 min of reaction time are not valid. Results are interpreted as follows:

**Reveal® Device Interpretation**

- All valid analyses (including negative controls) MUST present with an upper band on the Reveal® test device. Absence of upper band indicates INVALID device results as this band represents an internal device control.
- The Reveal® positive control samples MUST display TWO bands (upper and lower) in the Reveal® test device. Absence of both bands indicates INVALID device results. Absence of the lower band indicates a failure of the positive control.
- Matrix Spike samples MUST display TWO bands (upper and lower) in the Reveal® test device. Absence of both bands indicates INVALID device results. Absence of the lower band indicates a possible matrix effect.
- Method Blank (negative control) samples MUST display ONE band (upper) in the Reveal® test device. Absence of upper band indicates INVALID device results. Presence of the lower band indicates contamination during sample processing.
- Unspiked test samples displaying two bands (upper and lower) in the Reveal® test device indicate an *E. coli* O157:H7 POSITIVE sample.
- Unspiked test samples displaying an upper band only in the Reveal® test device indicate an *E. coli* O157:H7 NEGATIVE sample.

## BAX® DETECTION PROCEDURE

### Sample Collection and Storage

For each water matrix being analyzed for the presence of *E. coli* O157:H7, collection of a 1 L grab sample is recommended. A 200 mL sub-sample is analyzed for each water matrix. It is recommended that at least 20% of the total number of samples analyzed for each matrix are subjected to matrix spike analyses. The sample volume used for each matrix spike should be equivalent to the volume analyzed for the test samples (i.e., 200 mL sub-sample). For finished water (tap water) samples, residual chlorine must be neutralized by adding sodium thiosulfate to a final concentration of 0.01%.

Between sample collection and analysis, it is recommended that all samples are stored in a refrigerator (4°C).

### Sample Concentration

1. Prepare the vacuum, vacuum manifold and Microfunnel™ disposable filter units. Ensure that the funnel portion of the unit is securely snapped into place as it can dislodge during shipment.
2. Flame sterilize forceps or tweezers (used for transferring membrane filters from manifold to the 50 mL conical tubes). The forceps or tweezers require sterilization between each sample membrane.

3. Using an incubator set at 42°C (or 35°C), warm a sufficient volume of sterile TSB for the total sample load. A minimum of 10 mL of TSB is required for each sample.
4. Retrieve the water samples from the refrigerator and mix thoroughly by inversion.
5. Dispense 200 mL of each test water sample into individual sterile bottles and label with unique identification codes and designate samples as unspiked test samples.
6. Prepare matrix spike and method blank according to procedures described in QC section.
7. Filter each sample (unspiked, matrix spike or method blank) through individual Microfunnel™ filters, at a flow rate of approximately 100 mL per min.
8. Remove each membrane filter from its filter housing by snapping off the funnel portion of the unit. Use forceps to remove the membranes and place in individual pre-labeled 50 mL conical centrifuge tubes. Sterilize the forceps between membranes. The filter funnel and base can be discarded.
9. Add 10 mL of 1X TSB to each conical tube, cap tube and vortex samples at maximum speed for 1 min. Allow to stand for at least 10 sec and repeat vortex (1 min).
10. For each sample, transfer the filter eluant to an individual pre-labeled Leighton tube. Discard the conical tube and filter membrane.
11. Continue with the sample enrichment and IMS procedure.

**Sample Enrichment and IMS**

12. Place a Dynal Sample Mixer in an incubator set at 42°C.
13. Attach Leighton tubes to the Sample Mixer. Verify that the caps on the Leighton tubes are on securely.
14. Start the Dynal Sample Mixer, adjust rotational speed to 24 rpm and incubate samples for 2 h.
15. Remove the Leighton tubes from the Sample Mixer and add 100 µL of Dynabeads® anti-*E. coli* O157:H7 to each Leighton tube. (Paramagnetic beads will settle during shipment. Thoroughly resuspend the beads before use).
16. Return Leighton tubes to the Sample Mixer and continue incubation at 42°C for one additional hour (rotational speed 24 rpm).
17. During the last hour of enrichment, pre-label Eppendorf tubes, set out magnets, and prepare IMS wash buffer.
18. Remove the Leighton tubes from the Sample Mixer one at a time and place in the MPC®-1 magnets.
19. Ensure flat side of the Leighton tube is flush with the magnetic portion of the MPC®-1.
20. Gently rock the MPC-1 and Leighton tube from end to end (180 degrees motion) for 3 min.
21. Decant the supernatant, holding the Leighton tube in the magnet, with the magnet side being uppermost (i.e., pour the supernatant away from the magnet). Paramagnetic beads should form a visible pellet on the flat side wall of the Leighton tube, adjacent to the magnet. When removing the cap from the Leighton tube, ensure that the tube does not move/rotate in the MPC®-1.
22. Exercise appropriate infectious material handling precautions in discarding the supernatant. Autoclave any liquid waste before disposal.

*1st Wash*

23. Remove the Leighton tube from the MPC®-1 and resuspend the bead-bacteria complex using 0.8 mL of IMS wash buffer using a 1 mL pipet.
24. Gently perform repeated pipeting to resuspend the beads. Transfer the resuspended beads to a 1.5 mL Eppendorf tube prelabeled with the sample ID.
25. Add an additional 0.5 mL of IMS wash buffer to the Leighton tube and gently rinse the inside of the Leighton tube. Add this wash to its respective Eppendorf tube.
26. Repeat steps 7-14 with all samples using the MPC®-1 magnet before continuing to step 16.
27. Place the Eppendorf tubes (containing 1.3 mL - 1.5 mL of the sample bead-bacteria complex) into MPC®-S magnets and slide the magnet strip into place. Align tubes such that the hinged side of the tube is flush with the magnet.
28. Gently rock the Eppendorf tubes in the MPC®-S from end to end (90 degree motion) for 3 min.
29. Keeping the Eppendorf tubes secured in the MPC®-S, gently open the tube caps and use a P1000 pipetman (or Pasteur pipet) to remove the supernatant. Avoid disturbing the bead-bacteria complex that has formed a pellet on the hinged side wall of the Eppendorf tube. Change pipet tips between samples to avoid cross-contamination. Discard the supernatant, exercising appropriate care for handling infectious material.

*2nd Wash*

30. Remove the magnetic strip from the MPC®-S (the tubes can remain in the holder). Add 1 mL of IMS wash buffer to the bead pellet in each tube, using a P1000 or 1 mL disposable pipet. Use a clean pipet tip (or disposable pipet) for each Eppendorf tube.
31. Cap all the Eppendorf tubes and gently rock from end to end (90 degrees motion) 5-10 times or until the bead pellet has been resuspended.
32. Insert the magnet into the MPC®-S and rock the entire assembly from end to end (90 degrees motion) for 3 min.
33. Keeping the Eppendorf tubes in the MPC®-S, gently open the tube caps and use a P1000 pipetman (or Pasteur pipet) to remove the supernatant. Avoid disturbing the bead-bacteria complex. Change pipet tips between samples to avoid cross-contamination. Discard the supernatant, exercising appropriate care for handling infectious material

*3rd Wash*

34. Remove the magnetic strip from the MPC®-S (the tubes can remain in the holder). Add 1 mL of IMS wash buffer to the bead pellet in each tube, using a P1000 or 1 mL disposable pipet. Use a clean pipet tip or disposable pipet for each Eppendorf tube.
35. Cap all the Eppendorf tubes and gently rock from end to end (90 degrees motion) 5-10 times or until the bead pellet has been resuspended.
36. Insert the magnet into the MPC®-S and rock the entire assembly from end to end (90 degrees motion) for 3 min.
37. Keeping the Eppendorf tubes in the MPC®-S, gently open the tube caps and use a P1000 pipetman (or Pasteur pipet) to remove the supernatant. Avoid disturbing the bead-bacteria

complex. Change pipet tips between samples to avoid cross-contamination. The supernatant is discarded, exercising appropriate care for handling infectious material

38. Remove the Eppendorf tubes from the MPC®-S assembly and transfer to a regular microtube rack.
39. Add 1 mL of IMS wash buffer to each tube and resuspend the bead-bacteria pellet by vortexing for 10-30 sec.
40. Process all samples to step 39 before continuing with the BAX® detection protocol.

## BAX® DETECTION

1. Set one heat block (or water bath) to 37°C and a second heat block (or water bath) to 95°C.
2. Prepare lysis reagent according to procedures in QC section.
3. Vortex each 1 mL sample concentrate for 10-30 seconds.
4. For each sample, transfer 500 μL of the sample into a clean pre-labeled 1.5 mL Eppendorf tube.
5. For each sample, store one of the two tubes at 4°C for later use, such as verification of BAX® positives.
6. For each sample, place the second tube in the MPC-S, aligning the hinge against the magnet.
7. Rock the MPC-S from end to end (90 degrees motion) for 3 min.
8. Remove the supernatant, and discarding appropriately.
9. Remove the magnet from the MPC-S.
10. Add 50 μL of lysis reagent to the bead pellet in each Eppendorf tube and vortex (10-30 seconds) to suspend the bead-bacteria complex. (The resuspended bead-bacteria complex can be transferred to BAX® lysis tubes (DuPont Qualicon, Wilmington, DE) or PCR tubes if required for fit into the laboratory heat blocks.)
11. Prepare the BAX® positive and negative controls according to procedures described in QC section.
12. Place all tubes (containing samples and controls) into a heat block or water bath at 37°C and incubate for 20 min.
13. Transfer the tubes from 37°C to a heat block (or water bath) set at a 95°C. Incubate for 10 min.
14. Place sample tubes in refrigerator (4°C) or a cold block to cool for 5-10 min.
15. Place the tubes containing the samples in the MPC-S with magnet in place. Ensure the hinged side of the tubes is adjacent to the magnet. Allow the MPC-S to stand on a bench top until the beads form a pellet against the side wall of the tube (3 min). (Positive and negative control tubes do not need to be separated using the MPC-S since they do not contain IMS beads.)
16. Place a BAX® PCR tube holding rack onto a pre-chilled (4°C) PCR cooling block. Insert the required number of BAX® PCR tube strips (each strip consists of eight PCR tubes) into the tube holding rack. Remove the caps from the BAX® PCR tubes. If an entire strip is not required the unneeded tubes can be cut off and stored at 4°C for future use.
17. For each sample, remove the supernatant from the bead pellets (approximately 50 μL) and add to the PCR tablet in an individual BAX® PCR tube. It is important to avoid carry-over of IMS beads.
18. Following hydration of the PCR tablet with 50 μL of the test samples or controls, close the tubes using the provided optical caps.

19. The hydrated BAX® PCR tubes can remain in the PCR cooling block up to 30 min or be placed in a refrigerator (at 4°C).
20. Turn on the power to the BAX® cycler/detector and launch the BAX® system application. The system will guide you through a series of steps confirming the proper operation of the BAX® instrument. Use the Rack Wizard prompts to define the sample wells in the rack file. Choose *E. coli* O157:H7 as the target organism from the drop-down menu.
21. Upon completion of the rack file set-up, warm up the cycler/detector. Select OPERATION>RUN FULL PROCESS in the menu bar or select the thermometer icon in the tool bar. The cycler/detector requires 5 min to warm up.
22. When the cycler/detector has reached its normal operating temperature, place the prepared BAX® PCR tubes from step 19 into the BAX® System sample block. Follow the prompts to complete this process.
23. Once the Cycling program has been initiated, the status of the process is displayed on the screen. The cycling/detection program takes approximately 3.5 h to complete. The system will signal a prompt when the program is complete.
24. Results can be displayed either in a colored well format or graphically. Printouts of the results can also be made.

## BAX® RESULT INTERPRETATION

- The results can be viewed in a well format.
- A green colored well indicates that a sample is negative for *E. coli* O157:H7.
- A red colored well indicates that a sample is positive for *E. coli* O157:H7.
- A yellow colored well containing a '?' indicates an indeterminate result. Typically this result is due to a failure of the internal control. The sample contained in this well will need to be re-analyzed. For further assistance contact the manufacturer.
- A yellow colored well containing a '?' and a red slash through the well indicates a system signal error. This sample will need to be re-analyzed. The manufacturer must be contacted concerning this malfunction of the system.
- Individual well results can be viewed graphically.
- All valid analyses (including negative controls) must present with a peak for the internal control. Absence of this internal control peak indicates indeterminate results.
- All positive results will present with a peak for the internal control as well as a peak for the *E. coli* O157:H7. The height of the peaks is not significant.
- All negative results will present with only a peak for the internal control.

# CHAPTER 5
# GENERAL DISCUSSION

The human pathogen *E. coli* O157:H7, which has gained public notoriety due to numerous foodborne outbreaks of human disease, especially following consumption of undercooked meat products, can also be transmitted by the waterborne route, particularly where there is absence of or insufficient water treatment and/or disinfection. Although numerous outbreaks have occurred in the US over the last 20 years, perhaps the most widely publicized outbreak occurred in Walkerton, Ontario and was directly responsible for the deaths of seven individuals. The occurrence of waterborne outbreaks has created the need for water suppliers, state and/or federal regulatory agencies to respond rapidly to contamination issues with this organism, which can cause severe complications in a proportion of the affected (i.e., HUS or HC). To achieve this, it is imperative that rapid and sensitive methods are available and routinely employed to better understand the epidemiology of this organism in environmental water samples. It is also important to identify whether seasonality exists or organism occurrence is impacted by activities within the catchment area. Currently, the USEPA regulates systems that operate at least 60 days per year, serve 25 people or more or have 15 or more service connections as public water systems under the Total Coliform Rule. The TCR requires routine monitoring of tap water for all public water systems, with the frequency and number of samples varying with the size of the system and the number of people that are served. Detection of total coliforms indicates that the system is contaminated or vulnerable to fecal contamination, and the tests can be performed by various procedures that have been documented in 20$^{th}$ Edition of Standard Methods for the Examination of Water and Wastewater (APHA *et al.,* 1998). Examples of such methods include multiple tube fermentation using Lauryl tryptose and Brilliant green lactose bile broth. For larger volumes of drinking water or natural waters (100 mL – 1L), the membrane filtration (MF) technique can be employed with Endo-type medium containing lactose for growth of colonies. Alternatively, chromogenic substrate based tests (i.e., Colilert® and others) may be used. Based on the methods currently employed in water utility laboratories for compliance monitoring for the TCR, it would be advantageous for utility laboratories to couple such existing methods with new *E. coli* O157:H7 methods, thus allowing fecal coliform positives to be rapidly screened for the presence of this human pathogen.

In the current investigations, usefulness of existing immunological and molecular methods, currently employed for detection of *E. coli* O157:H7 in food samples, was examined for detecting this organism in environmental samples. Selected methods were examined either as stand alone procedures or when coupled with existing TC methods. The *E. coli* O157:H7 method was expected to serve as a preliminary screening tool and needed to be specific and sufficiently sensitive to detect low levels of contamination in water samples. Most importantly, the method had to be capable of delivering results within a single working day to allow the water utility to employ contingency strategies to safe guard consumer health.

The time frame allotted for this study did not permit development of novel procedures for detection of *E. coli* O157:H7. As a result the sensitivity, specificity, and reproducibility of five immunological test kits, utilizing the lateral diffusion format, and a single molecular assay was assessed using spiking studies with both target and non-target organisms. The Reveal® and BAX® assay were optimized into an over-all procedure incorporating the use of membrane filtration to concentrate samples. The concentrates were enriched in TSB and *E. coli* O157:H7 were simultaneous captured using immunomagnetic beads. These procedures, while sufficiently sensitive for environmental monitoring, are intended to serve as preliminary tools for determining *E. coli* O157:H7. As a caveat it is recommended that approximately 5%-50% of IMS concentrate is retained for subsequent analysis using an alternative confirmatory tool where samples are determined to be presumptively positive for *E. coli* O157:H7. A number of different approaches can be used for *E. coli* O157:H7 confirmation and include slow fermentation of sorbitol for phenotypic characterization of *E. coli* O157:H7 on Sorbitol MacConkey agar (sMAC). Other improved formulations such as Cefixime-SMAC, Sefixime-Rhamanose (CR-sMAC) and Cefixime-Tellurite (CT-sMAC) have been used, as well as development of selective, differential media (i.e., Rainbow Agar) that enable *E. coli* identification following chromogenic substrate metabolism. In conjunction with phenotypic characteristics, biochemical profiles, using API 20E, have also enabled confirmation up to the species level. Serological assays, utilizing antibodies (polyclonal or monoclonal) targeting either O157 lipoloysaccharide or H7 flagellar antigen or Shiga toxin (1 and 2), have been used in one or more immunological detection formats, such as ELISA's or dipstick immunoassays, and rapid methods, such as latex agglutination procedures, to provide information on serotyping. Characterization of Shiga toxins (1 and 2) by determining their cytotoxic effects on monolayers of Vero cells (African green monkey kidney cells) in the presence or absence of specific neutralizing antibodies is considered the gold standard for determining the production of toxins; however, these procedures are cumbersome.

During this project, a selective media (CHROMagar™) as well as two different latex agglutination tests were examined for their usefulness in rapid confirmation. These limited data generated in this study indicated that results needed to be interpreted with caution. For example API 20E was not able to generate reproducible results following repeat analysis of a single sample, the CHROMagar™ exerted an inhibitory effect on certain strains of *E. coli* O157:H7 and, while the latex agglutination tests worked effectively, large quantities of ($>1x10^7$) of organisms were required to generate a positive reaction. Additionally immunological procedures may be prone to cross reactivity issues. While detection of *E. coli* O157:H7 in multiplex PCR formats, targeting virulence genes such as *Stx* and *eaeA,* may be a more specific and sensitive option, *E. coli* serotypes other than O157:H7 can produce verotoxins. The use of restriction site specific (RSS)-PCR, which utilizes primers homologous to specific restriction enzyme sequences and amplifies genomic DNA segments between restriction sites, was used to create genetic profiles for various strains of *E. coli* O157:H7 (Kimura *et al*., 2000). Although this assay demonstrated the same profile for *E. coli* O157 and *E. coli* O55:H7, it demonstrated a better correlation with a monoclonal antibody based ELISA procedure than polyclonal antibody based latex agglutination tests. The RSS-PCR procedure may serve as alternative, more convenient confirmatory tools by generating 'fingerprint' patterns, which are distinct for different *E. coli* serotypes.

In conclusion, rapid isolation of low levels (0.04 – 0.2 CFU per mL) of *E. coli* O157:H7 from 200 mL samples of source or finished water samples was achieved by membrane filtration, followed by enrichment and simultaneous IMS capture and detection by either an immunological (Reveal®) or molecular (BAX®) assay. Sensitivity and reproducibility trials in a single laboratory validation were confirmed using 'blind' multi-laboratory round robin validation. Current data suggest it would be prudent to use the BAX® assay only for analysis of raw water samples, whereas both the BAX® and Reveal assay could serve as preliminary screening tools for analysis of finished water samples. Cost analysis based on consumables only indicated the overall procedures were inexpensive (Reveal® costs $23.82 per sample processed by the overall procedure and BAX® costs $24.42 per sample processed by the overall procedure). The round robin participants, which included water utility laboratories, required little or no training to successfully complete analysis within a single day, and in their feedback, indicated the methods were convenient and user-friendly.

Based upon the research conducted in this project, water utilities will be able to adopt a two-prong approach for detection of *E. coli* O157:H7. When following a total coliform test, or *E. coli* analysis, and suspecting a health risk from *E. coli* O157:H7 in the sample, methods described here could be applied to rapidly ascertain information on the presence of *E. coli* O157:H7 directly in a sub-sample of the total coliform test. Alternatively, a water utility suspecting *E. coli* O157:H7 contamination in a water supply, as a result of increased incidents of disease within the served communities, could apply these methods to test directly for the organism. Both approaches will be of great value to water utilities as this information will allow employment of rapid contingency measures to ensure continued safety to public health.

Future research should consider the procedures developed in this project to determine the relative cross reactivity of the Reveal® and BAX® assays using raw water samples taken from various sources. In addition, the occurrence of *E. coli* O157:H7 in source and finished water samples and information on seasonal distribution needs to be determined using geographically distinct sources.

# REFERENCES


Ackman, M., S. Marks, P. Mack, M. Caldwell, T. Root and G. Birkhead. 1997. Swimming-associated haemorrhagic colitis due to *Escherichia coli* O157:H7 infection: evidence of prolonged contamination of fresh water lake. *Epidemiol. Infect.* 119(1):1-8.

Akashi, S., K., Joh, A.Tsuji, H. Ito, H. Hoshi, T. Hayakawa, J. Ihara, T. Abe, M. Hatori, T. Mori and T. Nakamura. 1994. A severe outbreak of haemorrhagic colitis and hemolytic uremic syndrome associated with *Escherichia coli* O157:H7 in Japan. *European J. Ped.* 153:650-655.

Albert, M., S. Faruque, A. Faruque, P. Neogi, M. Ansaruzzaman, N. Bhuiyan, K. Alam and M. Akbar. 1995. Controlled study of *Escherichia coli* diarrheal infections in Bangladeshi children. *J. Clin. Microbiol.* 33:973-977.

APHA, AWWA and WEF (American Public Health Association, American Water Works Association and Water Environment Federation). 1998. *Standard Methods for the Examination of Water and Wastewater.* 20th Edition. Washington, D. C.: APHA.

Barwick, R., D. Levy, G. Craun, M. Beach and R. Calderon. 2000. Surveillance for Waterborne-Disease Outbreaks- United States, 1997-1998. In: CDC Surveillance Summaries. *MMWR* 49 (No. SS-4): 1-35.

Belongia, E., M. Osterholm, J. Soler, D. Ammend, J. Braun and K. Macdonald. 1993. Transmission of *Escherichia coli* O157:H7 infection in Minnesota child day-care facilities. *JAMA* 269:883-888.

Bitzen, M., K. Ludwig, M. Klemt, H. Konig, J. Buren and D.E. Muller-Wiefel. 1993. The role of *Escherichia coli* O157 infections in the classical (enteropathic) haemolytic uremeic syndrome: results of a Central European, multicenter study. *Epidemiol. Infect.* 110:183-196.

Bockemuhl, J., S. Aleksic and H. Karch. 1992. Serological and biochemical properties of Shiga-like toxin (Verocytotoxin)-producing strains of *Escherichia coli*, other than O-group 157, from patients in Germany. *Zentralbl. Bakteriol.* 276:189-195.

Boyce, T., D. Swerdlow and P. Griffin. 1995. Current concepts: *Escherichia coli* O157:H7 and the hemolytic-uremic syndrome. *N. Engl. J. Med.* 333:364-368.

Brewster, D., M. Browne, D. Robertson, G. Houghton, J. Bimson and J. Sharp. 1994. An outbreak of *Escherichia coli* O157 associated with a children's paddling pool. *Epidemiol. Infect.* 112:441-447.

Bukhari, Z., R.M. McCuin, C.R. Fricker, and J.L. Clancy. 1998. Immunomagnetic separation of *Cryptosporidium parvum* from source water and samples of various turbidities. *Appl. Environ. Microbiol.* 64(11):4495-4499.

Canada Communicable Disease Report. 2000. Waterborne outbreak of gastroenteritis associated with a contaminated municipal water supply, Walkerton, Ontario, May-June 2000. Available <http://www.phac-aspc.gc.ca/publicat/ccdr-rmtc/00vol26/dr2620eb.html> (cited November 2003)

Cebula, T., W. Payne and P. Feng. 1995. Simultaneous identification of strains of *Escherichia coli* serotype O157:H7 and their Shiga-like toxin type by mismatch amplification mutation assay-multiplex PCR. *J. Clin. Microbiol.* 33:248-250.

Chapman, P., C. Siddons, P. Zadik and L. Jewes. 1991. An improved selective medium for the isolation of *Escherichia coli* O157. *J. Med. Microbiol.* 35:107-110.

Coia, J. 1998. Clinical, microbiological and epidemiological aspects of *Escherichia coli* O157 infection. *FEMS Immuno Med Microbiol* 20:1-9

Cransberg, K., J. van den Kerkhof, J. Banffer, C. Stijnen, K. Wernars, N. van de Kar, J. Nauta and E. Wolff. 1996. Four cases of hemolytic uremic syndrome- source contaminated water? *Clin. Nephrology*. 46:45-49.

Cravioto, A., R. Reyes, F. Trujillo, F. Uribe, A. Navarro, J. de la Roca, J. Hernandez, G. Perez and V. Vazquez. 1990. Risk of diarrhea during the first year of life associated with initial and subsequent colonization by specific enteropathogens. *Am. J. Epidemiol.* 131:886-904.

Dev, V., M. Main and I. Gould. 1991. Waterborne outbreak of *Escherichia coli* O157. *Lancet*. 337:1412.

Doyle, M. and J. Schoeni. 1984. Survival and growth characteristics of *Escherichia coli* associated with hemorrhagic colitis. *Appl. Environ. Microbiol.* 48:855-856.

Farmer, J., III and B. Davis. 1985. H7 antiserum-sorbitol fermentation medium: a single-tube screening method for detecting *Escherichia coli* O157:H7 associated with hemorrhagic colitis. *J. Clin. Microbiol.* 22:620-625.

Farnum, A. D., S. Gandhi, M. A. Jensen, W. M. Barbour, N. Kesty, S. Y. Tseng, G. Tice, M. E. Schaffer, A. M. Stoltzfus, B. W. Anadaloro, W. D. Hudson, A. M. Hochberg and J. Hazel. 2000. Evaluation of an internal PCR control reaction for the BAX® screening system assays for foodborne pathogens. Presented at ASM General Meeting, Los Angeles, CA, May 21-May 25, 2000.

Feng, P. 1993. Identification of *Escherichia coli* O157:H7 by DNA probe specific for an allele of *uid*A gene. *Mol. Cell Probes*. 7:151-154.

Fratamico, P., R. Buchanan and P. Cooke. 1993. Virulence of an *Escherichia coli* O157:H7 sorbitol-positive mutant. *Appl. Environ. Microbiol.* 59:4245-4252.

Gilbert, L., and P. Blake. 1998. Outbreak of *Escherichia coli* O157:H7 infections associated with a waterpark. *Georgia Epidemiol Report* 14:1-2.

Gunzer, F., H. Bohm, H. Russman, M. Bitzan, S. Aleksic and H. Karch. 1992. Molecular detection of sorbitol-fermenting *Escherichia coli* O157 in patients with hemolytic-uremic syndrome. *J. Clin. Microbiol.* 30:1807-1810.

Hildebrand, J., H. Maguire, R. Holliman and E. Kangesu. 1996. An outbreak of *Escherichia coli* O157 infection linked to paddling pools. *Comm. Dis.* 6:R33-R36.

Hrudey, S. E., P. Payment, P. M. Huck, R. W. Gillham and E. J. Hrudey. 2003. A fatal waterborne disease epidemic in Walkerton, Ontario: comparison with other waterborne outbreaks in the developed world. *Water Sci Tech*. 47:7-14.

Ibekwe, M., P. Watt, C. Grieve, V. Sharma and S. Lyons. 2002. Multiplex flurogenic real-time PCR for detection and quantification of *Escherichia coli* O157:H7 in dairy wastewater wetlands. *Appl Environ Microbiol*. 68:4853-4862.

Isaacson, M., P. Canter, P. Effler, L. Arntzen, P. Bomans and R. Heenan. 1993. Haemorrhagic colitis epidemic in Africa. *Lancet*. 341:961.

Johnson, J., C. Brooke and S. Fritschel. 1998. Comparison of BAX™ for screening *E. coli* O157:H7 vs. conventional methods for detection of extremely low levels of *Escherichia coli* O157:H7 in ground beef. *Appl Environ Microbiol*. 64:4390-4395.

Jones, I. and M. Roworth. 1996. An outbreak of *Escherichia coli* O157 and Campylobacteriosis associated with contamination of a drinking water supply. *Pub. Health*. 110:277-282.

Karch, H., J. Russmann, H. Schmidt, A. Schwarzkopf and J. Heesemann. 1995. Long-term shedding and clonal turnover of enterohemorrhagic *Escherichia coli* O157 in diarrheal disease. *J Clin Microbiol*. 33:1602-1605.

Karmali, M. A., M. Petric, C. Lim, P. C. Fleming and B. T. Steele. 1983a. *Escherichia coli* cytotoxin, haemolytic-uraemic syndrome, and haemorrhagic colitis. *Lancet* ii:1299-1300.

Karmali, M. A., B. T. Steele, M. Petric and C. Lim. 1983b. Sporadic cases of haemolytic uremic syndrome associated with faecal cytotoxin and cytotoxin producing *Escherichia coli* in stools. *Lancet* i:619-620.

Keene, W. E., J. M. McAnulty, F.C. Hoesly, L. P. Williams, K. Hedberg, G. L. Oxman, T. J. Barrett, M. A. Pfaller and D. W. Fleming. 1994. A swimming-associated outbreak of hemorrhagic colitis caused by *Escherichia coli* O157:H7 and *Shigella sonnei*. *N. Eng. J. Med.* 331:579-584.

Kehl, S. C. 2002. Role of the laboratory in the diagnosis of enterohemorrhagic *Escherichia coli* infections. *J. Clin. Microbiol.* 40:2711-2715.

Kimura, R., R. Mandrell, J. Galland, D. Hyatt and L. Riley. 2000. Restriction-site-specific PCR as a rapid test to detect enterohemorrhagic *Escherichia coli* O157:H7 strain in environmental samples. *Appl Environ Microbiol* 66(6): 2513-2519.

Lee, S., D. Levy, G. Craun, M. Beach and R. Calderon. 2002. Surveillance for Waterborne-Disease Outbreaks- United States, 1999-2000. In: CDC Surveillance Summaries. *MMWR* 51 (No. SS-8): 1-43.

Levy, D., M. Bens, G. Craun, R. Calderon and B. Herwaldt. 1998. Surveillance for Waterborne-Disease Outbreaks- United States, 1995-1996. In: CDC Surveillance Summaries. *MMWR* 47 (No. SS-5): 1-33.

Li, W. and M. Drake. 2001. Development of a quantitative competitive PCR assay for detection and quantification of *Escherichia coli* O157:H7 cells. *Appl. Environ. Microbiol.* 67:3291-3294

March, S. and S. Ratnam. 1986. Sorbitol-MacConkey medium for detection of *Escherichia coli* O157:H7 associated with hemorrhagic colitis. *J. Clin. Microbiol.* 23:869-872.

Mariani-Kurkdjian, P., E. Denamur, A. Milon, B. Picard, H. Cave, N. Lambert-Zechovsky, C. Loirat, P. Goullet, P. Sansonetti and J. Elion. 1993. Identification of *Escherichia coli* O103:H2 as a potential agent of hemolytic-uremic syndrome in France. *J. Clin. Microbiol.* 31:296-301.

Mead, P. and Griffin, P. 1998. *Escherichia coli* O157:H7. *Lancet* 352:1207-1212

Morgan, D., C. Newman, D. Hutchinson, A. Walker, B. Rowe and F. Majid. 1993. Verotoxin producing *Escherichia coli* O157 infections associated with the consumption of yogurt. *Epidemiol. Infect.* 111:181-187.

Nataro, J. and J. Kaper. 1998. Diarrheagenic *Escherichia coli*. *Clinical Micobiology Reviews.* 11:142-201

New York Department of Heath. 2000. Health commissioner releases *E. coli* outbreak report. Available <http://www.health.state.ny.us/press/releases/2000/ecoli.htm> (cited November 2003).

O'Conner, D. R. 2002. Report of Walkerton Inquiry: Part 1- The events of May 2000 and related issues. Available <http://www.attorneygeneral.jus.gov.on.ca/english/about/pubs/walkerton/part1/> (cited November 2003).

Ochoa, T. and T. Cleary. 2003. Epidemiology and spectrum of disease of *Escherichia coli* O157. *Curr. Opin. Infect. Dis.* 16:259-263

Olsen, S. J., G. Miller, T. Breuer, M. Kennedy, C. Higgins, J. Walford, G. McKee, K. Fox, W. Bibb and P. Mead. 2002. A waterborne outbreak of *Escherichia coli* O157:H7 infections and hemolytic uremic syndrome: implications for rural water systems. *Emerg. Infect. Dis.* 8:370-375.

Padhye, N. and M. Doyle. 1992. *Escherichia coli* O157:H7: epidemiology, pathogenesis, and methods for detection in food. *J. Food Prot.* 55:555-565.

Paunio, M., R. Pebody, M. Keskimaki, M. Kokki, P. Ruutu, S. Oininen, V. Vuotari, A. Siitonen, E. Lahti and P. Leinikki. 1999. Swimming-associated outbreak *of Escherichia coli* O157:H7. *Epidemiol. Infect.* 122:1-5.

Riley, L. W., R. S. Remis, S. D. Helgerson, H. B. McGee, J. G. Wells, B. R. Davis, R. J. Hebert, E. S. Olcott, L. M. Johnson, N. T. Hargrett, P. A. Blake and M. L. Cohen. 1983. Hemorrhagic colitis associated with a rare *Escherichia coli* serotype. *N. Engl. J. Med.* 308:681-685.

Schoeni, J. and M. Doyle. 1994. Variable colonization of chickens perorally inoculated with *Escherichia coli* O157:H7 and subsequent contamination of eggs. *Appl. Environ. Microbiol.* 60:2958-2962.

Shelton, D. and J. Karns. 2001. Quantitative detection of *Escherichia coli* O157 in surface waters by using immunomagnetic electrochemiluminescence. *Appl. Environ. Microbiol.* 67:2908-2915.

Swerdlow, D. L., B. A. Woodruff, R. C. Brady, P. M. Griffin, S. Tippen, H. D. Donnell, E. Geldreich, B. J. Payne, A. Meyer and J. G. Wells. 1992. A waterborne outbreak in Missouri of *Escherichia coli* O157:H7 associated with bloody diarrhea and death. *Ann. Intern. Med.* 117:812-819.

Tarr, P. I., M. A. Neill, C. R. Clausen, S. L. Watkins, D. L. Christie and R. O. Hickman. 1990. *Escherichia coli* O157:H7 and the hemolytic uremic syndrome: importance of early cultures in establishing the etiology. *J. Infect. Dis.* 162:553-556.

Tice, G., J. C. Rohrbeck, S. D. Dabell, D. H. Brunk, D. J. Lee, A. M. Stoltzfus, M. E. Schaffer and S. Gandhi. 2000. Comparison of a system for the homogeneous detection of PCR products with a gel-based system. Presented at ASM General Meeting, Los Angeles, CA, May 21-May 25, 2000.

Tozzi, A., A. Niccolini, A. Caprioli, I. Luzzi, G. Montini, G. Zacchello, A. Gianviti, F. Principato and G. Rizzoni. 1994. A community outbreak of haemolytic-uraemic syndrome in children occurring in a large area of Northern Italy over a period of several months. *Epidemiol. Infect.* 113:209-219.

U. S. Environmental Protection Agency. 1989. Total Coliform Rule (TCR). *Fed. Reg.* 54 (124):27544-27568.

Wang, G. and Doyle, M. P. 1998. Survival of enterohemorrhagic *Escherichia coli* O157:H7 in water. *J Food Prot.* 61(6):662-667.

Warrner, M., K. Kuo, L. Williams, B. Adam, C. Langkop, R. Ruden, B. Francis and T. Haupt. 1996. Lake-associated outbreak of *E. coli* O157:H7 in Illinois. *MMWR.* 45:437-379.

Weir, E. 2000. *Escherichia coli* O157:H7. *CMAJ* 163 (2):205

Whittam, T. and R. Wilson. 1988. Genetic relationships among pathogenic *Escherichia coli* of serogroup O157. *Infect. Immun.* 56:2467-2473.

Whittam, T. S., M. L. Wolfe, I. K. Wachsmuth, F. Orskov, I. Orskov and R. A. Wilson. 1993. Clonal relationships among *Escherichia coli* strains that cause hemorrhagic colitis and infantile diarrhea. *Infect. Immun.* 61:1619-1629.

Willshaw, G., H. Smith, D. Roberts, J. Thirwell, T. Cheasty and B. Rowe. 1993. Examination of raw beef products for the presence of Vero cytotoxin producing *Escherichia coli,* particularly those of serogroup O157. *J. Appl. Bacteriol.* 75:420-426.

Willshaw, G. A., J. Thirwell, A. P. Jones, S. Parry, R. L. Salmon and M. Hickey. 1994. Vero cytoxin-producing *Escherichia coli* O157 in beefburgers linked to an outbreak of diarrhea, haemorrhagic colitis and haemolytic uraemic syndrome in Britain. *Lett. Appl. Microbiol.* 19:304-307.

Wright, D. J., P. A. Chapman and C. A. Siddons. 1994. Immunomagnetic separation as a sensitive method for isolating *Escherichia coli* O157 from food samples. *Epidemiol. Infect*. 113:31-39.

# ABBREVIATIONS

| | |
|---|---|
| ATCC | American Type Culture Collection |
| AOAC | Association of Official Analytical Chemists |
| AWWA | American Water Works Association |
| AwwaRF | Awwa Research Foundation |
| | |
| BGBB | Brilliant Green Bile Broth |
| | |
| °C | degrees Celsius |
| CDC | Centers for Disease Control |
| CFU | colony forming units |
| CR-sMAC | cefixime-rhamanose sorbital-MacConkey agar |
| CT-sMAC | cefixime-tellurite sorbital-MacConkey agar |
| | |
| DNA | deoxyribonucleic acid |
| dsDNA | double stranded deoxyribonucleic acid |
| | |
| EAEC | enteroaggregative *E. coli* |
| *E. coli* | *Escherichia coli* |
| EHEC | enterohemorrhagic *E. coli* |
| EIEC | enteroinvasive *E. coli* |
| ELISA | enzyme linked immunosorbent assay |
| EPEC | enteropathogenic *E. coli* |
| ETEC | enterotoxigenic *E. coli* |
| | |
| FSIS | United States Department of Agriculture Food Safety Inspection Service |
| | |
| Gb3 | globotriaosylceramide |
| | |
| h | hour |
| HC | hemorrhagic colitis |
| HUS | hemolytic uremic syndrome |
| | |
| IMS | immunomagnetic separation |
| INPC | internal positive control |
| | |
| L | liter |
| LEE | locus of enterocyte effacement |
| LTB | Lauryl Trytose Broth |
| | |
| MCL | Maximum Contaminant Level |
| MDa | megadalton |
| MF | membrane filtration |
| min | minute |
| mL | milliliter |
| MQ | Milli-Q Ultrapure |
| MUG | methyl-umbelliferyl β-D-glucuronide |

| | |
|---|---|
| µL | microliter |
| µm | micrometer |
| | |
| NA | not applicable or not examined |
| | |
| P/A | presence/absence |
| PAC | project advisory committee |
| PBS | Phosphate Buffered Saline |
| PCR | polymerase chain reaction |
| PHLS | Public Health Laboratory Services |
| | |
| RAPD | random amplified polymorphic DNA |
| RSS-PCR | restriction site specific polymerase chain reaction |
| RNA | ribonucleic acid |
| rRNA | ribosomal ribonucleic acid |
| | |
| SDWA | Safe Drinking Water Act |
| sec | second |
| sMAC | sorbital-MacConkey agar |
| STEC | Shiga toxin producing *E. coli* |
| Stx1 | Shiga toxin 1 |
| Stx2 | Shiga toxin 2 |
| | |
| TC | Total Coliform |
| TCR | Total Coliform Rule |
| TSA | Tryptic Soy Agar |
| TSB | Tryptic Soy Broth |
| TTP | thrombotic thrombocytopenic purpura |
| | |
| UK | United Kingdom |
| USEPA | United States Environmental Protection Agency |
| | |
| VTEC | Verocytotoxin producing *E. coli* |

IWA Publishing's authorised EU representative for General Product Safety Regulations is Diane D'Arras, 15 rue Duret, 75116 Paris, France, e-mail: safety@iwap.co.uk.

Printed and bound by CPI Group (UK) Ltd, Croydon, CR0 4YY
17/09/2026
02218813-0011